The Big Book of
Makerspace Projects

About the Authors

Colleen Graves is a teacher librarian obsessed with learning common transformations, maker education, makerspaces and making stuff. Colleen brings a passionate artistic energy to the school library world; this passion earned her the School Library Journal/Scholastic School Librarian of the Year Co-Finalist Award in 2014 and Library Journal Mover and Shaker Innovator Award in 2016. She is an active speaker and presenter on makerspaces and the maker movement on a national level.

Aaron Graves is a school librarian with 18 years of experience in education. He is a mash-up of robot geek, book lover, and tech wizard. Aaron gained his perseverance for projects through collaborative and interactive art experiences as a member of the Good/Bad Art Collective. He is an active speaker and presenter on libraries, makerspaces, and research skills. In his free time he can be found writing, restoring microcars, or inventing something that makes you smile!

The Big Book of Makerspace Projects: Inspiring Makers to Experiment, Create, and Learn

Colleen Graves

Aaron Graves

New York   Chicago   San Francisco   Athens   London   Madrid
Mexico City   Milan   New Delhi   Singapore   Sydney   Toronto

Cataloging-in-Publication Data is on file with the Library of Congress

McGraw-Hill Education books are available at special quantity discounts to use as premiums and sales promotions or for use in corporate training programs. To contact a representative, please visit the Contact Us page at www.mhprofessional.com.

The Big Book of Makerspace Projects: Inspiring Makers to Experiment, Create, and Learn

2 3 4 5 6 7 8 9   ROV   21 20 19 18 17 16

ISBN   978-1-25-964425-2
MHID   1-25-964425-1

This book is printed on acid-free paper.

Sponsoring Editor
Michael McCabe

Editorial Supervisor
Donna M. Martone

Production Supervisor
Lynn M. Messina

Acquisitions Coordinator
Lauren Rogers

Project Manager
Patricia Wallenburg, TypeWriting

Copy Editor
James Madru

Proofreader
Claire Splan

Indexer
Claire Splan

Art Director, Cover
Jeff Weeks

Composition
TypeWriting

To my mom, Susan Price, for teaching me
that all materials can be re-used
for new purposes and for giving me
a strong sense of independence and initiative.
Colleen

To Glenn Graves and C.F. Ellis
who secretly taught and inspired me
under the guise of being an "assistant."
Thanks to you both
for sharing your skills
and demonstrating your craftsmanship
and ingenuity countless times!
Aaron

Contents

Acknowledgments

THIS BOOK COULD never have been finished without all the awesome makers who loaned us a listening ear when we: ran into trouble, couldn't get past the next step, or just needed a sounding board.

Thank you, Jay Silver and the entire Makey Makey team for the little Invention Kit that gave me (Colleen) the creative confidence to create enough projects to write a whole book!

We are forever grateful to Josh Burker for his support, encouragement, and sensible wisdom.

Without Jeff Branson as her "maker therapist," Colleen never would've gained her debugging super powers. Thanks for always asking, "How can I help?"

A big thank you to Bev Ball and Jie Qi for their encouragement and enthusiasm about our projects! (And for being inspirational in showing us how art and engineering go hand in hand!)

Circuit gurus Marshall Garth Thompson and Shane Culp were instrumental in always making sure we were on the right path (taking the correct "route" if you will!)

We are forever indebted to "Sensei" Trey Ford of the Denton Public Library for his patience in writing Arduino code with Colleen.

Thanks to Tim Sanchez for sharing his knowledge of physics and expertise in teaching it to the masses.

Val and Viv, thanks for making stuff with us and also giving us time to write. We love you.

Thank you fellow authors Chris Barton and Jeff Zentner for your support in our journey of becoming authors.

Thank you Michael McCabe and the whole McGraw Hill team (and Patty!) for asking the impossible of us! We never knew we could write a book. Your edits, revisions, and vision have made this such an exciting project!

Lastly, our deepest gratitude is to you dear reader. We hope all of these projects instill you with the creative confidence to experiment, create, and learn new things. We can't wait to see all of the great stuff you make!

CHAPTER 1

An Introduction to Making and Tinkering

WE CREATED THIS BOOK as a maker's handbook. The projects in this book will guide you through many "makerspace" fundamentals. Even if you have no experience in making, you'll be able to pick up this book and make over half the projects. Once you've completed those, you'll be an advanced maker and be able to complete the other half! For advanced makers who pick up this book, you'll find some hard fun in over half the projects right from the start! Plus, you'll enjoy helping others make the beginner projects to help us all in spreading the love for the maker movement.

But just to be clear, we don't want to encourage cookie-cutter making. We want you and other makers to create and make these projects together and then use all the knowledge you gain as building blocks for designing and making your own awesome creations!

Anyone can make things and follow instructions, but the real goal of a maker in a makerspace is to learn to tinker and tinker to learn. We want you to take these ideas and these projects and make them your own. Tinkering is a skill and a habit. It's something you did as a small child and can do again with just a little work! For those who struggle with tinkering, we've built in challenges for each project to get you thinking beyond the project and help you to take your work further. But don't stop there, tinker with the projects and go beyond the instructions—see how far you can stretch your

thinking. This is one of the reasons we've added a big challenge at the end of each chapter. Use your newfound creative confidence to take the ideas you've learned throughout the chapter and invent something new.

We want to build your skill set and fill your maker toolbox to the point where you feel comfortable designing, making, tinkering, and creating your own projects. We can't wait to see what you create! We are so excited to see, hear, and know about all the projects you make out of this book. Even more so, we want to see the amazing things you make after you master the basic concepts you've learned from making the projects in our book. We want to see your work, your ideas, and your inventions because now you are part of the *maker movement*. An integral part of making is learning something and then giving back to the community by teaching others to make. You'll see this happen in your own makerspace as your fellow makers gain creative confidence.

Sharing

This is why we are including our Twitter and Instagram handles. Please feel free to reach out to us and share your projects with the global maker community. Sharing your work and providing inspiration for others are important ways you can contribute to makers worldwide. Use the #bigmakerbook hashtag to share your work. We created a community page at colleengraves.org/bigmakerbook just to showcase *you*!

Making Projects Accessible

Makers have been making some of the projects in this book for years. We do not claim ownership of the original concept.

For those projects, we researched and reviewed multiple tutorials and then tinkered with the design to make them accessible for makers of any expertise level. For example, we knew we wanted a project that was easy enough for makerspace facilitators to help students build their own guitar. We wanted you to feel comfortable leading this type of workshop with your students. When we looked into making guitars with our own students, we found hundreds of cigar box guitar templates out there. However, many of them were too complicated or too time consuming to replicate in our own makerspaces. So Aaron spent time reviewing, blending, and tinkering until he created a version of a one string guitar project we felt anyone could make! We tested this project out on our own students and found it to be quite successful and fun. In fact, for the last few years, many of the projects in this book were tested in our library classrooms.

Classroom Tips

This is why we've included classroom tips in each chapter to help makerspace facilitators work with large groups of makers. We tried to avoid cookie-cutter projects that all end with the same product. A good maker project provides room for innovation, creativity, and uniqueness!

Each of these projects could be done as a class in which students tinker to learn concepts and learn to tinker simultaneously! As a makerspace facilitator, you might feel tempted to print out guides or worksheets listing concepts and detailed steps. Instead, let your students learn by playing and creating. The goal is to provide just enough instruction and guidance to give students basic skills and then let them take charge of their own learning and discovery. You need to set up an environment in which innovation is encouraged and students are allowed to find several ways to take on or master a challenge.

When hosting a makerspace workshop, the key is to have enough supplies and plan for troubleshooting. As much as possible, we've tried to troubleshoot these projects, but if you find yourself in a pickle that we haven't described … that's a good thing! Troubleshooting for errors will increase your perseverance and maker grit, which are qualities our students desperately need. In fact, troubleshooting (also known as *debugging*) is one of the highest maker superpowers, and frequently, making those mistakes will lead you to invention and learning something new.

Go Make Something!

Our main goal in our library makerspaces is for making and invention to be accessible to all our students. You are now one of our students. Go forth and *make*!

A Note On Safety

It's easy to get the maker fever, forget about safety, and rush through finishing a project. We strongly encourage you to take your time and always use the right tools because it's your responsibility to work safely. If you are working with young children, always stress the importance of safety. While we do encourage young people to make, please make sure that you supervise young children, wear the proper safety gear, and train students to use the tools for each project. We find that following these guidelines, our projects and our students' projects turn out better because we work through them slowly and with care. Remember to take breaks sometimes because this disciplined way of working can lead you to more "Aha!" moments and inventive ideas. P.S.: Always remember to turn off your soldering iron!

Starting Small and Low Cost

THESE STARTER PROJECTS were handpicked as beginner makerspace projects that anyone can do regardless of skill level. Plus, makers should be able to find these everyday materials around the house.

Project 1:	Brush Bot Warriors
Project 2:	Cardboard Arenas
Project 3:	Do-It-Yourself (DIY) Paperclip Switch for Brush Bot or Scribble Bot
Project 4:	Scribble Machines
Project 5:	Perfect Circle Machines with littleBits
Project 6:	Balloon Monorail
Project 7:	Balloon Hovercraft
Project 8:	Balloon-Propelled Car/Boat

Chapter 2 Challenge

Make something that moves!

Project 1: Brush Bot Warriors

Most of these robots you can make with materials you already have at your house. You may have seen a simple bristle bot before; it's time to take that idea to the next level with the brush bot.

Cost: Free–$

Make time: 10–15 minutes

Supplies:

Materials	Description	Source
Dollar-store supplies	Bristled brushes, electric toothbrush, rubber bands, erasers, AA batteries, masking tape, pencil erasers or glue sticks to create an eccentric weight	Dollar Tree
Battery holder	Single AA battery holder	Radio Shack SparkFun
Test leads	Alligator test leads or wires	Radio Shack SparkFun
Insulators and connectors	Heat shrink, small twist-on wire connectors, electrical tape, masking tape	Radio Shack Hardware store

Step 1: Get the Motor

Break open an electronic toothbrush or handheld dollar store fan to get to the motor for your brush bot (Figure 2-1). If you use a toothbrush motor, it will already be unbalanced. Look carefully at the shaft coming out of the end of the motor. Does it have a weight on the end that is slightly off center? If it does, the motor has an eccentric weight, which means it's scientifically ready to make a robot (Figure 2-2). If you have a fan motor, however, you'll have to

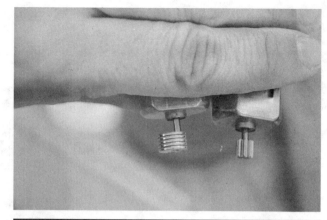

Figure 2-2   Eccentric weight (*left*). Geared and balanced motor (*right*).

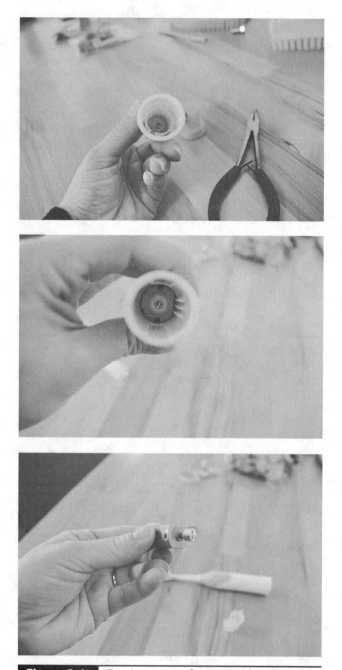

Figure 2-1   Freeing motor from toothbrush.

add an eccentric weight to the end of the motor to throw this little guy off balance.

Why do we want to throw it off balance? Because the unbalanced vibration is what makes your robot move! Technically, you can make your brush bot with the fan without even taking the motor out. You could simply break a blade off or add a small weight like a paperclip to one

fan blade, use a lot of tape to cover your weight (because you don't want it flying off, poking you in the eye!), and attach your entire fan to your scrub brush. But where is the fun in that? So get out some pliers and free that little motor from the fan.

Classroom tip: If you are making these bots with a class or large group of people, you'll probably want to free your motors beforehand, but keep one complete toothbrush on hand to show your makers how you can find resources for electronic projects in everyday things. You'll also want to point out the difference in a motor with or without an eccentric weight. This is a good way to teach your makers about how an eccentric weight offsets the mass of your gadget and that this vibration is what allows your bot to go!

Step 2: Add an Eccentric Weight

If your motor does not have an eccentric weight or has a straight shaft, you'll need to add an eccentric weight to the end of the shaft that will not interfere with rotation of the motor. A small piece of a glue stick works great, but you could also use the end of a pencil eraser as an eccentric weight. This weight should be just off center and provide enough vibration to get your bot moving. The vibration caused by the off-center weight is what disrupts your bot's static inertia and makes it travel across the floor.

The other disruption to the bot's inertia is the direction of the bristles and your placement of the battery pack and motor. If you plop your motor down in the center of your brush, what do you think will happen? How about if you place it near the end? This is where the tinkering fun really begins.

Step 3: Attach Motor and Battery Pack

Since we are trying to use everyday materials, you'll want to use something you can find around your house. Rubber bands actually work quite well for this project, but if you have access to a glue gun, you could attach your motor this way, but then you wouldn't be able to experiment with different placements. So use a rubber band to find the best placement to get the best movement on your brush bot before you break out your hot glue gun. While researching for this book, I found a great hack from Exploratorium's site where they note that at the Tech Innovation Museum, people used Velcro as a fastener for brush bot motors so that makers could tinker with placement in a quick and easy way.

Classroom tip: When sharing basic electronic projects with students, it is good to teach them a about battery safety. It is important to teach students that a circuit is a closed loop around which electrons flow. The battery is the source of those electrons. In the robots you are creating, those electrons are used to power motors when the circuit is complete. This means that you should never hook up the positive side of a battery to the negative side of a battery with an alligator clip because this will create a short circuit. This causes a large amount of energy to flow in a short time, creating the battery to overheat or, worse, creating a spark. It is essential that after projects are made, all batteries should be disconnected and stored properly.

Step 4: Clip and Strip Alligator Clips

I find that for this project, since we are not using a switch, it is easier to start and stop your bot simply by clipping the alligator clips to the connections on the motor. So we'll need to prepare your alligator clip by cutting it in half with scissors. Then take some wire strippers to expose the copper wire at each end where you snipped the clip in half. Lightly twist the newly exposed copper wires together. Use a wire connector to cover the exposed copper wires (Figure 2-3).

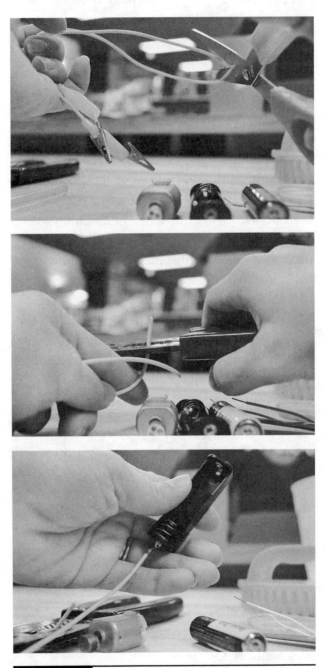

Figure 2-3 Clipping and stripping alligator clips.

Step 5: Hook Up Your Bot

It's time to hook up your bot! However, we don't want it moving before you are ready, so we'll make a simple DIY switch with duct tape or masking tape. Simply fold a piece of duct tape or masking tape in half and place it between the battery and the positive end of the battery pack. All you really need is something that is not conductive that you can place between the battery and the connection so that you can turn your bot on when you are ready (Figure 2-4). Now hook up your battery wires to the exposed copper wire on the alligator clips. You'll need one for each side. It actually doesn't matter which end goes to which side. Loop the little piece of wire from your battery holder onto the copper and lightly twist the wire so that it will stay on the clipped end of the alligator clip. Do the same thing on the other side. To keep exposed wires down to a minimum, you may wish to use heat shrink, or if your battery pack has wires already attached, you can use "standard wire connector" caps (pictured in Figure 2-3) to twist your exposed wires together. Then use the alligator clips to clip onto each tiny copper connection on the motor. In most cases, how you hook up your leads will determine the rotation of the motor.

Figure 2-4   Duct tape switch.

Classroom tip: Makers often ask if they will be shocked by the battery when hooking it up for the first time. With AA and AAA batteries, there is very little risk of shock because our skin is a relatively poor conductor. Although electricity flows through the battery into your body if you touch both terminals, there is not enough amperage for you to even notice.

Step 6: Drive Your Bot

It's time! Place your bot with other bots in your cardboard arena (steps in next project), and pull your DIY tape switch and let your robot rip (Figure 2-5)! If you notice that your robot is not moving quickly or spinning wildly in circles, push on the bristles to change your robot's path, or experiment with trimming the bristles (Figure 2-6). If you still don't like the way it is "driving," try moving the motor and/or battery pack. The most forceful brush bot wins!

Challenges

- What happens if you mount two motors to your brush bot powered by one battery? Now try two batteries.

- What happens when you trim the bristles at a greater angle?

- If you switch the leads on your motor, how does it affect your bot? Which way does it perform the best?

- If you have a brush that is square on one end, how can you change it so that it does not get stuck against the wall?

- Experiment with different brush styles. What works best?

There are more great ideas on brush bots at http://tinkering.exploratorium.edu/.

Figure 2-5 Brush bot assembled!

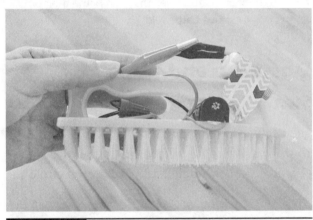

Figure 2-6 Trimming bristles.

Project 2: Cardboard Arena

Cost: Free–$

Make time: 15–30 minutes

Supplies:

Materials	Description	Source
Recyclables	Large corrugated cardboard box and chopsticks	Recycling bin
Office supplies	Duct tape or packing tape	Office supply store
Tools	Box cutter, ruler, marker	Hardware store

Step 1: Find a Good Box

Find a large area cardboard box that is long and wide but narrow in width. We are mainly interested in having a large flat piece of cardboard as a smooth surface for your bots to battle on. You only need about a 3-inch-high side, so you can cut the top off and use any remaining cardboard to cut 3-inch strips. These strips will become the channels for your brush bots (Figure 2-7).

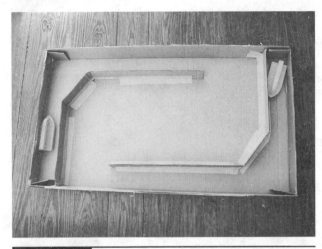

Figure 2-7 Long and wide cardboard arena.

Step 2: Use Tape

Reinforce the corners with duct tape, if needed. Create a channel by duct taping a piece of trimmed cardboard about 4 inches from the side of the wall (wide enough so that your bot won't turn completely around but instead be forced to travel forward because of the vibration of the motor). The track needs to be parallel to the edge of the board and should create a 90-degree angle (Figure 2-8).

Step 3: Make a Turn

About three-quarters of the way across the top, create an obtuse angle so that your bot will turn (Figure 2-9).

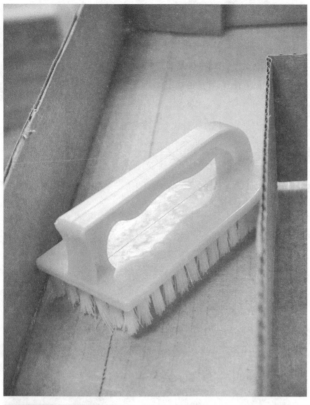

Figure 2-8 Ensure that width is determined by brush bot size.

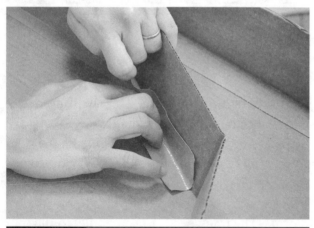

Figure 2-9 Taping channels and creating an obtuse angle.

Step 4: Diagonal Reflection

Create a second channel using diagonal reflection on the opposite side of your box so that your bot has an identical pathway to the battlefield.

Practice Run

- Battle your bots in a practice run. What are their natural pathways?

- Create your traps based on these pathways by putting traps in spots where the bot would not normally travel. (So the other bot will have to push it into the trap.)

Classroom tip: Before making cardboard arenas, let students see how brush bots behave by watching and tinkering with their driving patterns. Once students understand the nature of the robot's path of motion, have students work in small groups of three to four makers to build arenas as a collaborative task. After building their arena together, they can battle their bots on this collaborative battlefield!

Obstacles and Traps

Step 1: Pit of Doom

Elevate your arena, and cut a hole in the bottom to trap wayward brush bots in the pit of doom (Figures 2-10 through 2-12). We drew a circle

Figure 2-10 Elevated arena.

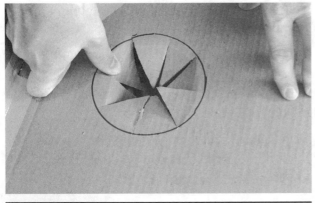

Figure 2-11 Pit of doom.

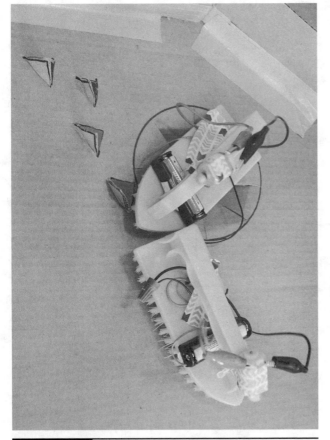

Figure 2-12 Brush bot trapped in pit.

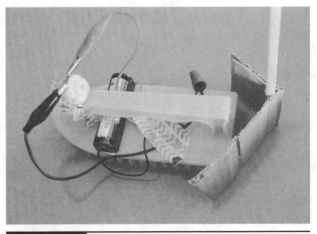

Figure 2-13 Brush bot trapped in wheel.

and cut the hole like you would slice a pizza, so that our brush bot will get trapped.

Step 2: Obstacles

Cut teeth in cardboard to stop your opponent's brush bot or even block it from going down into the pit of doom. Craft a trap wheel to spin brush bots or capture them. Use a chopstick, and tape an L-shaped piece of cardboard. Use a pencil to poke the hole for the chopstick, and make sure to do this through two boxes so that your trap wheel is stable (Figure 2-13).

Challenges

- See if you can create a battle arena with two obtuse angles and two acute angles.
- Create traps with items you have around your makerspace. What kind of trap can you invent with recycled goods?
- Can you make a jousting arena where the brush bots are forced to travel a certain path?
- Could you make a cardboard maze for your bots?

Project 3: DIY Paper Clip and Brad Switches

Cost: Free–$

Make time: 5–10 minutes

Supplies:

Materials	Description	Source
Recyclables	Reuse plastic from fruit containers, cardboard	Recycling bin
Office supplies	Metal paperclips, copper brads	Office supply store
Switch	Reuse a switch from a toothbrush or another project	Junk drawer

Steps for a Paperclip Switch

Step 1: Cut a small rectangle out of plastic (you could use cardboard, but remember, we are working with electricity, so plastic is a safer working material).

Step 2: Slice two small spots to push your copper brads through.

Step 3: Push copper brads through plastic, and catch one end of your paper clip. Remember that electricity likes to flow through the shortest avenue possible, so you need to make sure that you hook your paper clip as in Figure 2-14.

Step 4: Wire one end of your brush bot as you normally would.

Step 5: With the end you haven't wired, hook the cable from the battery to the DIY switch by hooking it under the copper brad that houses the paperclip (Figure 2-15).

Step 6: Take your stripped alligator clip and hook it under the opposite brad that is not touching the paper clip (Figure 2-16).

Step 7: Secure with tape, and clip your alligator clip to the motor. Your brush bot will now work only when the paperclip connects with the brad (Figure 2-17).

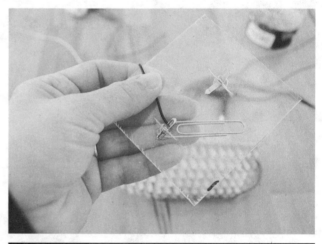

Figure 2-15   Attaching the other end of the battery to DIY switch.

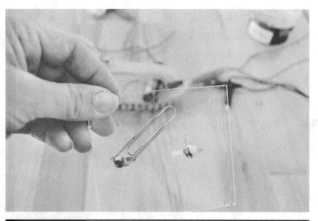

Figure 2-14   DIY paper clip switch.

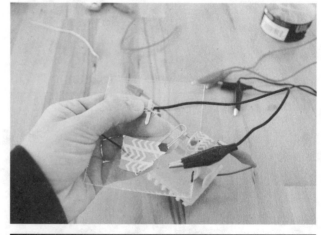

Figure 2-16   Attach the stripped end of alligator clip to the switch.

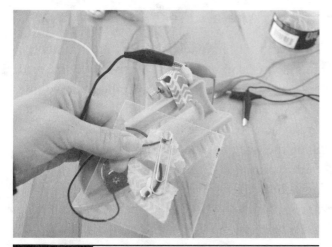

Figure 2-17 Fully wired for DIY switch.

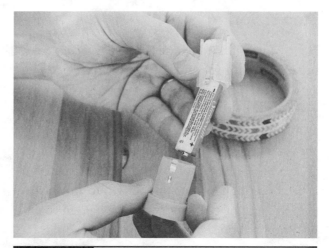

Figure 2-18 Switch assembly.

Reusing an Existing Switch

Since your toothbrush came with a switch, why not use it? Follow the instructions on this simple hack to use a preexisting switch for your brush bot.

Materials	Description	Source
Toothbrush parts	Spring, base of switch, battery, clear battery holder	Dollar Tree
Office supplies	Tape, rubber bands, pencils	Office supply store
Tools	Pliers	Hardware store

Step 1: Check Switch Assembly

To get started, place the base and battery holder together as they were in the body of the toothbrush (Figure 2-18). Make sure that the metal leads on the side slide together and are making contact.

Step 2: Secure Connections

Secure the metal leads that were held in place by the toothbrush body with a rubber band or tape. These metal leads travel up the length of the base switch and battery holder to supply power from the positive side of the battery (Figures 2-19 and 2-20).

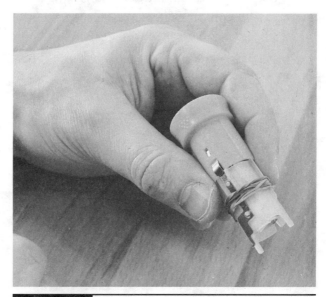

Figure 2-19 Secure connections with rubber band.

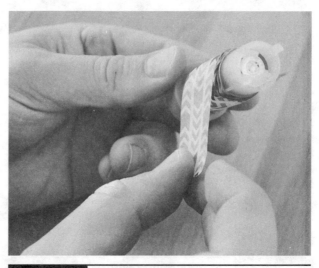

Figure 2-20 More secure connections with tape.

Step 3: Spring

Next, you will need to use the spring that was between the motor and the battery to create a lead for the negative side of the battery. This spring is often stretched out when you pull the motor out of the toothbrush; you can use a chopstick or a pencil to twist it back into shape (Figure 2-21). It does not need to be perfect, but it needs enough tension and length to touch the negative end of the battery and be held in place by a rubber band. You may need to wrap the rubber band more than once from top to bottom of the battery pack, and try to avoid getting in the way of the on/off switch.

Step 4: Wire and Test

Now your battery pack is ready to test. Attach an alligator clip test lead to the positive side (red in this example) and another to the negative side (black in the example in Figure 2-22). Now you can attach both these test wires to the motor. Slide the switch at the bottom of the battery pack on and see if the motor turns. If it does not, check the connection between the spring and battery and also check where the two metal leads meet on the side of the battery pack (Figure 2-23).

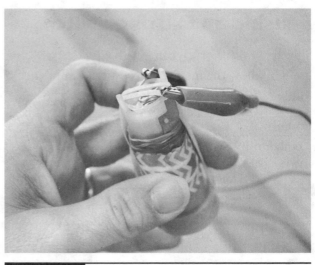

Figure 2-22 Attach alligator clips to switch.

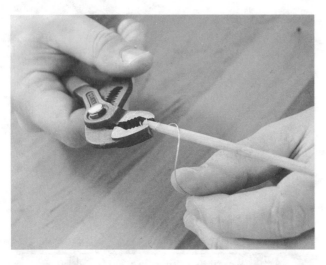

Figure 2-21 Respring your spring.

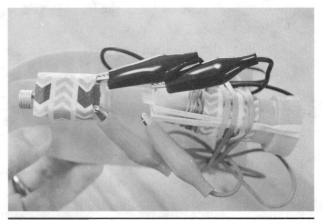

Figure 2-23 Full wiring for brush bot and reused switch.

Projects 4 and 5: Scribble Machines

Drawing is fun, but now that you know how to make a simple robot, let's put it to work. These two scribble machines are made with simple motors and markers.

Project 4: Scribble Bots

Once you get tired of your brush bot, you can reuse all your materials to make a robot that will do all your drawing for you! Well, okay, maybe not all your drawing, but this scribble bot is a hit with all ages.

Cost: $

Make time: 10–15 minutes

Supplies:

Materials	Description	Source
Dollar-store supplies	Plastic cups or containers, tape, markers, rubber bands, electric toothbrush, AA batteries	Dollar Tree
Recyclables	Plastic cups, reusable containers, markers	Recycling bin
Battery holder	Single AA battery holder	Radio Shack SparkFun
Test leads	Alligator clips or wires	Radio Shack SparkFun
Insulators (optional)	Heat shrink, small twist-on wire connectors, electrical tape, masking tape	Radio Shack Hardware store

Step 1: Give Your Bot Drawing Legs

Choose a cup or container and attach three to four markers with tape to the inside or outside. Ensure that the legs are attached firmly to the cup so that they will transfer the vibration from the motor. If they are loose, they will come off, and your robot won't move.

Step 2: Attach Motor and Power

Place a motor on the top of the cup so that the eccentric weight is hanging over the edge, and secure it with tape (Figure 2-24). However, don't use too much tape because you may want to move your motor to get better movement. You will need to place a battery in the battery pack and attach it with tape to the remaining space on the top of the cup. If your container allows, I like to use a roll of tape to secure the battery pack to the top of the container. Once you get a placement you like, you can also hot glue your motor and battery pack.

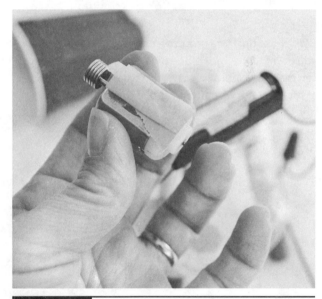

Figure 2-24 Attach motor with tape.

Step 3: Wire Up and Make Art

Cut an alligator clip in half, and strip about ¼ inch of the wire off the cut end (or reuse your clips from your brush bot). Use wire nuts to join the wires coming from the battery to the alligator clips. When you are ready to start making art, place a large piece of paper on a table or on the floor, and connect the alligator clips to the leads on the motor (Figures 2-25 and 2-26). (You can also add DIY switches from the

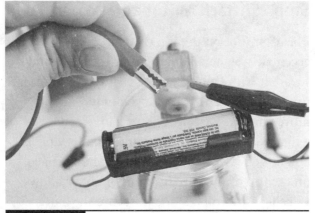

Figure 2-25   Attach alligator clips to motor.

earlier project so that you will have more power over your art bot as in Figure 2-27!)

Classroom tip: For a group of makers, make sure that you cover tables with butcher paper so that your groups can make collaborative artwork! Young makers may need assistance with wire stripping, or if you are doing this

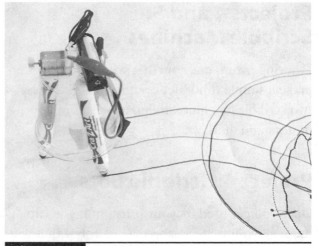

Figure 2-26   Scribble bot makes art!

project with a large group, this might be something to take care of in advance if time is limited. Remember that if you skip a step or have students who are too young, be certain to show them how you stripped the wire and the tool that you used.

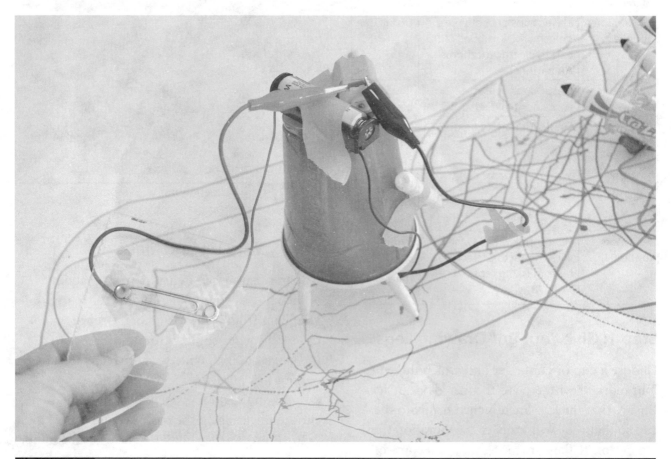

Figure 2-27   Add a DIY switch.

Challenges

- Can you make your bot draw straight lines?

- How many markers can you add to your bot before it stops moving?

- What can you do to your art bot to make it draw dotted lines?

- Add a marker or chopstick to your bot and joust with another bot. The first bot to fall over or stop moving loses! (See Figure 2-28.)

- For sumo wrestling bots, draw a large circle, and place bots in the center facing each other. The first bot to be pushed or to wander out of the circle loses.

- For tug-of-war bots, determine the direction of travel for each bot, and place an alligator clip on the back of each bot. Put a piece of tape on the center of the alligator clip, and draw a starting line. Draw two parallel finishing lines 8 inches away from the starting line. Turn both bots on, and the first to pull the other across the line wins.

- What kind of switch can you add to your scribble bot? Consider adding a littleBits remote trigger (as described in the next project) so that you can control your robot by remote! FYI: You'll have to wire up a littleBits motor and add an eccentric weight.

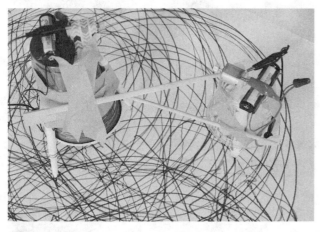

Figure 2-28 Jousting scribble bots.

Project 5: Perfect Circle Machine with littleBits

Cost: $$

Make time: 10–15 minutes

Supplies:

Materials	Description	Source
littleBits Gizmos and Gadgets Kit or Pro Library	Power (p1), remote trigger (i7), wire (w1), dc motor (o5)*	littleBits.cc
45 rpm record and adapter	Do not use a new record or one that you treasure.	Garage sale Thrift store
Small zip ties	6- to 8-inch zip ties	Hardware store
Remote	Remote for TV, projector, etc.	Living room
Tools	Marker, ruler	School or office supply store

*Optional. If you want to use different types of buttons to control your circle machine instead of a remote, you can turn it on and off with these supplies: power for your second circuit, button (i3), slide dimmer (i5), infrared LED (o7), etc.

Step 1: Test Your Circuit

This circuit is composed of the battery power supply, i7 remote trigger, w1 wire, and o5 dc motor with Motor Mate adapter. After you connect your littleBits, test the circuit with an infrared remote. The power supply needs to be on first, but then you can press any button on your TV or DVD player remote to trigger the i7 infrared remote trigger and make the motor move! When you press and hold any button on the remote, it should tell the trigger to send power to your motor and keep your motor on (Figure 2-29).

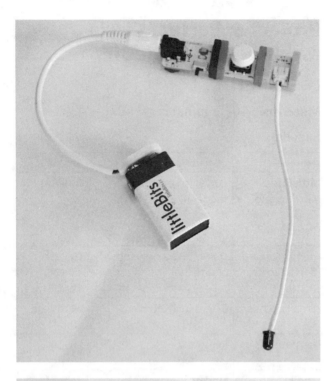

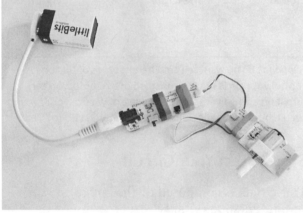

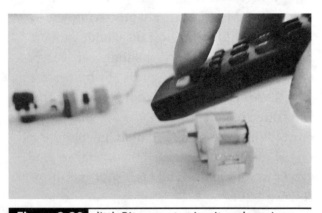

Figure 2-29 littleBits remote circuit and receiver circuit.

Step 2: Attach Your Bits to a Ruler

Cut a small piece of cardboard, and place it underneath the o5 dc motor between the legs of the bit to keep the legs from bending into the middle. Align the motor shaft with the 7-inch mark (or about an inch away from the center if you are using a longer ruler). It is important that you have your ruler facing up and your zip tie heads facing toward the table because our perfect circle will be rotating around on one of these zip tie heads. Use two zip ties to attach the motor bit to the ruler. With the zip tie heads at the base of the ruler, trim the first zip tie about ¼ inch away from where it locks into place. This will create a foot for the circle machine to rotate on (Figure 2-30). Trim the second zip tie about ⅛ inch away from where it connects so that it will not interfere with the rotation. Now attach a w1 wire to the motor bit, and connect the other end to the i7 remote trigger if using this circuit. Finally, attach the power supply, and plug in the battery and battery cable. We placed the power bit on top of the battery and used a single rubber band to hold it in place around the 9½-inch mark on the ruler.

Figure 2-30 Ruler rests on zip tie foot.

Step 3: Put a Record On

Push a 45 rpm record onto the MotorMate accessory, and place the MotorMate onto the littleBits dc motor. In our tests, we found that records that do not need a 45 rpm adapter work best. If you have a record that uses a plastic 45 rpm adapter, use tape to secure it to the MotorMate. Now lean the ruler so that the record touches the surface of the paper and rests on the zip tie foot.

Step 4: Marker and Balance Test

Attach a large Expo marker toward the end of the ruler that doesn't have your circuit. This will counterbalance the battery and circuit on the opposite end of the ruler. Adjust the marker position so that it has enough weight to just touch the paper surface. You can also adjust your circuit and battery position to achieve the perfect balance. You may need to return to this step to get your bot running in perfect circles (Figure 2-31).

Figure 2-31 Finished project.

Step 5: Push the Button

Now it's time for the moment of truth. Push the button on the remote, and watch your bot start drawing circles. It may take a few nudges for the circle machine to run smoothly, so give it a gentle tap to see if the record is ready to spin. If your bot does not move, revisit step 2 to check your zip-tie balance and step 4 to adjust your marker and circuit; make adjustments to balance as necessary.

Classroom tip: For large groups of makers, consider hosting an art-bot challenge with your littleBits and see what other kinds of drawing robots your students can invent! Can students find another way to make a perfect circle bot? What if they wire littleBits up like our scribble bot? Why won't that invention make perfect circles?

Challenges

- Remove the remote and use the infrared LED bit to test out different input buttons (slide switch, dimmer, etc.).

- Change the type of paper or use cardboard under the circle machine. How does it affect the performance of the bot?

- Switch out the 12-inch ruler for an 18-inch ruler or a yardstick. How large of a circle can you draw? How does moving the marker affect your perfect circle?

- How many markers can you add? What is the difference in diameter versus location on the ruler?

- What can you change to create a dotted line for your circle?

Projects 6–8: Balloons and Straws

We have all seen someone try to blow up and tie a balloon, lose their grip, and then watch the balloon shoot wildly across the room as the air inside is forced out the small opening. What if we could harness the power of the air pressure in the balloon and make a monorail or hovercraft or even propel a toy car? These next few projects are all based on *Newton's third law of motion*: "For every action, there is always an opposite and equal reaction." To harness the thrust (or reaction principle) created by the air, we need to force it through an even smaller opening or nozzle. For most of these builds, you will use a simple bendy straw to focus your balloon's power. To get started, try a simple experiment. Use a rubber band to fix a balloon on the short end of a bendy straw. Blow up the balloon and release it. Compare it to a balloon with no straw. How are they different? Now complete the next few projects to harness your new balloon knowledge!

Project 6: Balloon Monorail

Cost: Free–$

Make time: 5–10 minutes

Supplies:

Materials	Description	Source
Dollar-store supplies	Bendy straws, assorted balloons, clear tape, assorted rubber bands	Dollar Tree
String	Nylon string, monofilament line	Craft store
Paper (optional)	For adding wings or experimental tinkering solutions	School or office supply store

Step 1: Attach the Bendy Straw to the Balloon

The first thing you will need to do is attach a balloon to a straw so that you can blow air into the balloon engine after you've attached it to a slack-line track. Take a bendy straw and cut 2 inches away from the bendable joint on the long side. Place the straw inside the balloon up to the bendy joint, and use a rubber band to secure it. Sometimes twisting the balloon around the straw helps to seal this connection (Figure 2-32). Test your connection by blowing up your balloon. If your balloon comes loose, wrap the rubber band tighter, and add some tape.

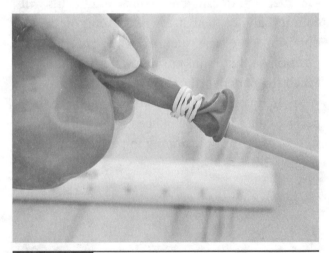

Figure 2-32 Get bendy with it.

Step 2: Attach the Balloon to the Track

Now it's time to get that air pressure under control and attach our balloon to a slack-line track to create a monorail. Trim off the long side of a bendy straw and center it on the spot where you attached your balloon to the first bendy straw (Figure 2-33). Wrap tape around the two straws, but be sure not to wrap the tape too tightly around the straws because it will keep your balloon from traveling. Air must be able

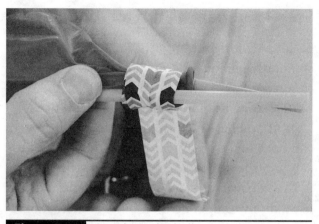

Figure 2-33 Attaching the balloon to the track.

Figure 2-34 Rev your engine!

to flow out of the first straw, and the first straw needs to be free to travel along the string track. You should use monofilament fishing line, nylon string, or a very smooth string for the track. To create a track, slide the string through the straight straw. Secure both sides of the string to the wall or a chair by tying them or using tape. Try to keep the string as level and taut as possible for now.

Classroom tip: Have two students per slack-line track. This way they'll already be set up to joust their balloon monorail cars. Make sure that each student has his or her own balloon car because you obviously only want to share ideas not germs!

Step 3: Test It Out

Now for the best part, back the balloon monorail car up to the start of the track, and blow through the first straw to rev your engine (Figure 2-34). Let go and see what happens! Does your balloon travel quickly or slowly? Does it move faster at the beginning of its journey or near the end? Don't just stop there. Let's try some of the following challenges to see what else you can learn.

Challenges

- How far can you make your balloon monorail go? Instead of attaching it to the wall, let a buddy hold the string, and see how far your balloon will go.

- Can your balloon travel uphill? Change the angle of your track, and see how it affects the travel speed.

- What happens when you attach two balloons to your monorail car? Can you joust with your balloons?

- Try adding wings (paper airplane) or something to make your car spin.

- What happens if you use different sizes of straws to hold your balloon to the track?

- What happens if you change the number of straws or the size of the straws used to put air into the balloon?

- Try using tape after you blow the balloon up to limit the amount of air coming out of the balloon. How does this change the performance?

Project 7: Balloon Hovercraft

Cost: Free–$

Make time: 5–10 minutes

Supplies (Figure 2-35):

Materials	Description	Source
Dollar-store supplies	Bendy straws, assorted balloons, assorted rubber bands	Dollar Tree
Craft-store supplies	Modeling clay, hot glue gun, hot glue sticks	Craft store
Lids	Small soda bottle lids around 1 inch Reuse lids from soda bottles, milk jugs water bottles that open and close by pushing/pulling	Recycling bin
CD or record	Old CDs or 45 rpm records	Grandpa's record collection

Step 1: Prepare a Path for Air Power

Select a lid for your project! We found that standard 1-gallon milk jug lids fit nicely into 45 rpm records and smaller water bottle or soda bottle lids work quite nicely for use with old CDs. If you have a lid from a water bottle that can open and close by pushing and pulling, you can turn the airflow on and off and skip step 2. If you have a standard lid, you will need to use a drill with a ¼-inch bit or an X-Acto knife to create a hole in the middle of the lid. Use a pair of pliers or a clamp to hold the lid in place instead of holding it in your hands (Figure 2-36).

Step 2: Position a Straw and Seal the Leaks

At this point you need to decide how long you want to keep this project. Plasticine works great but will eventually dry out and cause air leaks. Hot glue is a longer-lasting solution, but it is

Figure 2-35 Record hovercraft supplies.

Figure 2-36 Drill a hole.

not available to everyone or every age level. Place a straw through the hole, and then use plasticine (Figure 2-37) or hot glue to seal the leaks between the straw and lid (Figure 2-38). Once you have your straw in place, trim it evenly with the bottom of the lid, and make sure that it extends about 2 inches above the top of the lid.

Step 3: Secure Balloon

Place the balloon over the straw, and use a rubber band to secure it. If you have a push-pull top, place the rubber band over the top plastic piece that moves up and down, and use a rubber band to secure it. Make sure that you can open and close it to control your airflow. Inflate your balloon to see if there are any leaks around where the balloon is secured to the straw or lid.

Step 4: Add a Hover Disk

Now it's time to attach your balloon to a CD or vinyl record base.

Figure 2-37 Plasticine.

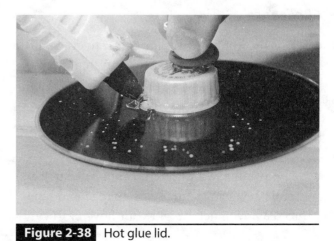

Figure 2-38 Hot glue lid.

- *Plasticine.* If you are using plasticine, roll out a thin coil that goes around the circumference of lid. Place the coil on the bottom of the lid, and press down. Make a larger coil that is slightly larger than the circumference of the lid, and press it into place around the outside of the lid.

- *Hot glue.* Place a bead of hot glue around the bottom of the lid. Center the lid on the record, and press down. Allow the glue to cool; then place a bead of glue around the edge of the lid to give this joint some added strength.

Step 5: Hover On!

Blow up the balloon, and place the CD or record on the floor, balloon side up. As the balloon starts to deflate, you will notice that the air is forced out over the flat surface of the CD. This creates just enough lift for the balloon to hover. Give your CD a little push, and watch it float across the floor like an air hockey puck (Figures 2-39 through 2-41).

Classroom tip: Hot glue use is fine for some students, but make sure that you cover the rules and expectations. Determine safety by age and maturity! You will definitely not want to use hot glue with K–4 students, but they will enjoy watching it work. Set up a glue station that you can monitor or control, and use the opportunity to talk about how a glue gun works and why it is dangerous.

Challenges

- This project only uses air power to create lift. How can you make a hovercraft that moves on its own?

- What is the largest disk you can make float with a standard balloon? Would a Frisbee or paper plate work the same?

- Can you make a hovercraft that will carry a load? Try using some spare change and tape to see how much weight your hovercraft can lift?

Figure 2-39   Plasticine-attached hovercraft.

Figure 2-40   Hot glued–attached hovercraft.

- How does changing the size of the balloon affect the performance of hovercraft?

- How about the surface? Does your hovercraft work the same on different surfaces? What is the best surface? Is this why hovercraft cars aren't mainstream?

Figure 2-41 Record hovercraft.

Project 8: Balloon Boats and Cars

Cost: Free–$

Make time: 10–15 minutes

Supplies (Figure 2-42):

Materials	Description	Source
Dollar-store supplies	Bendy straws, assorted balloons, assorted rubber bands, tape	Dollar Tree
Craft-store supplies	Modeling clay, hot glue gun, hot glue sticks	Craft store
Wheels	Lids from water or soda bottles and CDs	Recycling bin
Car and boat bodies	Plastic food trays with no holes, cardboard, a variety of water and soda bottles	Recycling bin
Axles	Chopsticks, pencils, skewers, toothpicks	Recycling bin
Play cars	Play cars that are lightweight (essential!)	Toy box
Tools	Hammer, nail, scrap block of wood	Hardware store

Figure 2-42 Balloon car/boat supplies.

Step 1: Balloon Power

Both of these projects are very similar to the hovercraft and monorail in the way they are powered with the recoil or reaction principle. You will need to use a rubber band to secure a balloon to the short end of a flexible straw as in the preceding projects. You should wait to trim your straw until you decide on what car or type of container you are going to use for your car or boat.

Step 2: Chassis for Car or Hull for Boat

Plastic bottles, small boxes, and just flat cardboard makes great bodies for balloon cars. You may decide to use an existing toy car; if so, try to locate one that is lightweight and has smooth wheels. You will also want to make sure that the car is big enough that you can tape or hot glue a straw onto it. For boats, a plastic bottle or a simple fruit or plastic sandwich container works great (Figure 2-43). If you are making a boat, skip to step 4.

Classroom tip: If you are working with small children and want to make boats, reuse sandwich containers or small trays so that students don't have to cut bottles in half. Save time and add safety.

Figure 2-43 Boat chassis.

Step 3: Rolling

If you are building a car from scratch, bottle lids, Ping-Pong balls, and jar lids make great wheels. Use a nail to punch a hole through the middle of the lids so that you can attach them to a round chopstick or skewer. These can then be placed through a straw to create an axle (Figure 2-44). You can also use small wooden spindles, but you may need to add tape to your chopstick axle so the wheel stays steady on the axle (Figure 2-45). To create your axle, cut two straws that are 1 inch longer than the width of your vehicle (Figure 2-46). Center your straws, and then tape one on the front third of your chassis and the other on the back. (This is something you can experiment with as you work through this project. However, if you put your axles too close, your wheels will touch.) Place one wheel on the

Figure 2-44   Wheels on chopstick/skewer axles.

Figure 2-45   Tape to reduce spinning.

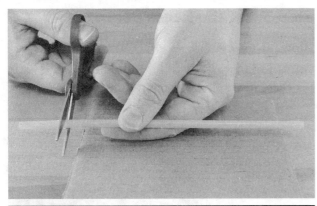

Figure 2-46 Cut straws a little bigger than the chassis.

Figure 2-47 Axles and wheels attached to chassis.

skewer, and then slide it through the straw to get an idea of how long the axle needs to be. Remove the axle from the straw, and then trim it to the desired length. Place it back through the straw, and then attach the other wheel. Repeat this step to install the rear axle and wheels (Figure 2-47).

Step 4: Add a Built-in Propulsion System

Because this car has no motor, you'll need to attach a balloon as your built-in propulsion system. Prepare your balloon as you have in other projects by inserting a straw and securing

air leaks with a rubber band. Check for leaks, and then tape the exposed straw to the car, making sure that you have the straw overhanging at the end of your car for thrust and for ease of blowing air into the balloon! This propulsion works because the air from the balloon is escaping at the back of your homemade vehicle.

For a boat, make sure that you have a place to tape down your balloon engine if needed (see cardboard platform in Figure 2-48), and then cut a small X where you want your straw to come out of your boat engine. You may not need a platform. See our soda bottle boat in Figure 2-49; the angle of the straw in the bottle actually held our engine steady. The most important thing is that the straw end will need to be in the water for your boat to work! The propulsion of air through

Figure 2-48 Balloon propulsion system car.

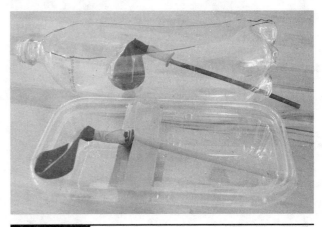

Figure 2-49 Balloon propulsion system boat.

the water is what makes the boat move. If you cut a hole in your boat, you need to reinforce the hole with modeling clay so that your boat doesn't spring a leak! Make sure you place this modeling clay inside your boat (Figure 2-50).

Step 5: Let the Races Begin

Blow air into your balloon, and watch your car or boat go (Figures 2-51 and 2-52)! This propulsion system works on the reaction

Figure 2-51   Balloon car in action.

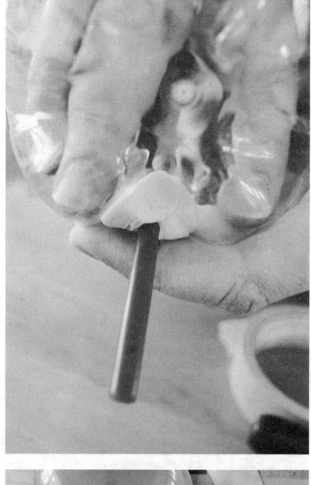

Figure 2-50   Seal the leaks!

Figure 2-52   Balloon boat in action.

principle. So, if your car doesn't move or doesn't move quickly, you need to think about the weight of your car and the ease of movement of your axle and wheels. If the thrust of the balloon is less than the weight of your car, you will lose every race. Can you lighten your materials so that the thrust is more powerful than the weight of the car? If the wheels of your car cannot spin freely, you need to problem solve and ease the friction.

Challenges

- What would happen if you attached more balloons?

- Could you add a dishwasher detergent nozzle to this project to control the airflow?

- Why does the car speed up at the end? Is there a way you could harness this thrust?

- What if you add a superlong straw? Will it make your car go faster or slower?

- Try this technique on your favorite Hot-Wheels car!

"Make Something That Moves" Challenge

Take all the ideas you learned in this chapter and come up with something new! Instead of using air to power your car, could you attach a motor? Could you use littleBits to motorize your recycled car invention? Or even use littleBits to motorize your cardboard arena?

What other items could you use to propel an object based on Newton's third law? Could you find ways to harness the power of rubber bands, carbonated soda, Mentos, springs, a fan, etc.?

Make something that moves and then take pictures of your challenge project. Tweet the pictures to us @gravescolleen or @gravesdotaaron or tag us on Instagram and include our hashtag #bigmakerbook to share your awesome creations. We will host a gallery of your projects on our webpage.

Smart Phone Projects

THESE FUN STARTER projects are for smart phones because tons of makers have them. It's a pretty easy way to add tech to cardboard and learn something interesting!

Project 9:	Pepper's Ghost
Project 10:	Smart Phone Hologram/Illusion Challenge
Project 11:	Smart Phone Projector

Chapter 3 Challenge

Smart phone stand challenge.

Project 9: Pepper's Ghost

Pepper's Ghost is a very old and very easy illusion to re-create in its simplest form. It relies on reflection and light to create a ghostly image. You've probably seen examples of it at a popular amusement park or at a live concert. With a few pieces of recycling material and your smart phone, you can have your own high/low-tech illusion in minutes!

Cost: Free–$

Make time: 10–15 minutes

Supplies:

Materials	Description	Source
CD case	Old CD case	Recycling
Small light	Flashlight, LED tea light, or candle	Junk drawer
Box	Cardboard box tall enough for case to stand in	Recycling
Scissors	—	School supplies
Tape	Clear tape	—
Black background	Black construction paper or cloth	Craft store
Marker	Black marker	Craft store
Craft or utility knife	—	Craft store

Step 1: Experiment with Illusions

The easiest way to start experimenting with creating an illusion is to use a clear CD case, black paper, clear tape, and a LED tea light or flashlight. You will also need a couple of small figurines or toys that are brightly colored and a piece of black construction paper or black fabric (Figure 3-1).

This illusion relies on reflection and light to create a ghostly image. You will need to find a room with medium to low light and a surface next to a wall. Position a large piece of black construction paper or cloth so that one half rests

Figure 3-1   Illusion experiment.

on the surface and the other rests against the wall. Use tape to secure your paper or fabric to the wall and table. Open the CD case about 45 degrees. Place the back of the case at a 90-degree angle away from the wall on the black paper or fabric. The clear front of the case should be open 45 degrees. Place your figurine about 6 to 10 inches away so that it appears to be looking at the front of the case. Now stand directly in front of your illusion facing the wall. You should see a ghostly reflection of your figurine in the front of the CD case.

Things to Try

- What happens when you move the figure further away?

- What happens if you shine a flashlight on your figurine?

- Add another figurine to the mix. Place it behind the front of the glass. How does it help your illusion?

- What happens if the viewer is not looking straight ahead but from the side?

Step 2: Control Your Viewer

Now that you are beginning to master the basics of this illusion, it is time to make it more convincing. Pepper's Ghost is often used in

amusement parks, theaters, and performance centers to make a ghostly reflection. Are you ready to re-create this effect on a smaller scale? In the preceding earlier experiment, the source of our ghost was exposed because you could see the figurine making the reflection. In this project, we are going to create a window for our viewer so that we do not expose our trick. This will allow us to control what the viewer is seeing and hide the figurine making the reflection. You will need a box that is at least 5 inches tall, 8 inches wide, and over 8 inches deep for this project.

To get started, let's create a window that is 2 inches wide and 2 inches tall. Choose the side on which you want to make your window, measure 1 inch away from the edge, and make a mark on the top and bottom of the box. Draw a line to connect the two marks.

To control what the viewer will see and hide the CD case, we want to make sure that our window is right in the middle of the CD case we are using as a reflective surface. Make a mark on the line 2.5 inches from the bottom of the box. This is the center of the CD case. Now place a mark 1 inch above and 1 inch below the center mark (Figure 3-2). These marks denote the top and bottom of the window.

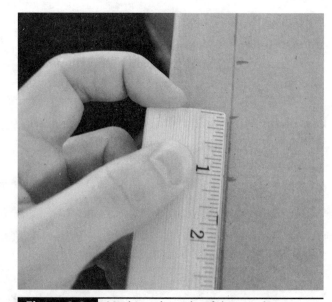

Figure 3-2   Marking the side of the window.

Measure across 2 inches, and draw a line parallel to the one you just made. Mark the top and bottom of the window, and draw a line to complete the square window (Figure 3-3). Use a craft knife or box cutter to cut the window out.

Figure 3-3 Marking the top, bottom, and right side of the window.

Step 3: Use Your Illusion

Place the back of your CD case against the wall to the right or left of the window. Use tape or hot glue to secure the case to the side of the box. Open the case to 45 degrees, and use a small piece of tape at the bottom to hold it in place (Figure 3-4).

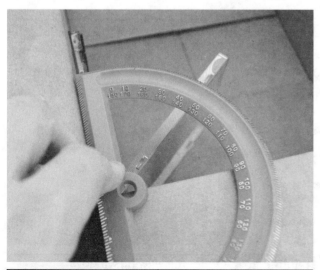

Figure 3-4 Placement of the CD case at 45 degrees.

Now place your figurine in the box so that it appears to be looking at its own reflection. It may be faint, but try to adjust the position of your figurine so that you can look through the window and see its reflection.

To hide the figurine and not reveal the illusion, the box must be closed, but before that is done, the figurine inside must be illuminated. It is good to place another figurine in the box near where the ghostly image will appear so that you can try to make it appear that the two are interacting (Figure 3-5).

Figure 3-5 Interactive figure.

The easiest way to illuminate the figure is to cut a hole in the box directly above the figurine. Another option is to cut a hole just large enough for a flashlight to shine down on the figurine (Figure 3-6). You can also use a tea light or

Figure 3-6 Illumination.

the long LED littleBit to light your ghost and figurine.

In the tradition of a parlor trick, this allows you to partially open the box and let viewers see the figurine but no ghost. When they look through the window, you can shine a light on the ghost figure, and suddenly it will appear as in Figure 3-7.

Figure 3-7 Complete illusion.

Step 4: Cover Your Tracks

To cover any signs of the CD case and to keep the CD case from picking up any reflections, cover the inside of the box with black or gray construction paper. You may also want to make a backdrop for your figurines.

Get Smart

Step 1: Smart Phone Image

Smart phones are a great way to imitate the way Pepper's Ghost is used in large-scale stage performances. The technique will still use a 45-degree angle to create the illusion, but the smart phone will supply the ghostly image. We

will need to position the CD case so as to make it easy for users to start and stop videos.

Optional Supplies	Description
Smart phone	For projecting your ghostly image
Straws	For supporting the CD case
Ghost	Figurines, small toys, or even record yourself with a black backdrop!
littleBits	Power (p1), 9-V battery, battery cable, forkBit (w7), 4 × wireBit (w1), long LEDBit (o2), 2 × pulseBit (i16), rgb LEDBit (o3), bargraphBit (o9), LEDBit (o1)

Step 2: Smart Phone Viewing

Begin by cutting a viewing hole. Find the center of the front of the box, and make a line. Make marks 1 inch to the left and right of the center line at the top and bottom of the box. Connect the marks to create vertical lines that will make the sides of your window. Make marks at 1.5 and 3.5 inches on both of the side lines. Connect those marks to create the top and bottom of the window, as in Figure 3-8. Cut the window out with a box cutter.

Use a protractor to open the case to a 45-degree angle. With the CD case open, measure and cut some straws that are about the

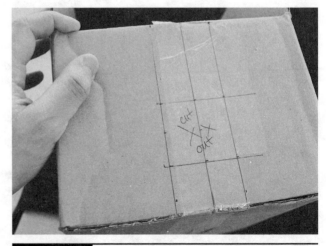

Figure 3-8 Center the viewing window.

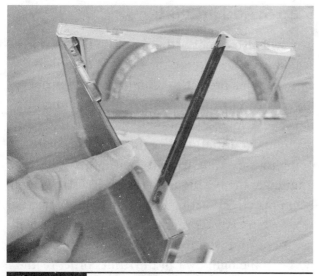

Figure 3-9 Taping straws in place.

same height as the lid. Use a permanent marker to color the straws black. Secure the position of the CD case at a 45-degree angle with the tape and straws, as in Figure 3-9.

Once you have the CD case secure, it's time to test the position of your CD case and determine where the hole for your phone needs to go. Place the back of the CD case on the floor of the box so that it opens away from the viewing window (Figure 3-10). You can add figurines behind the open case to interact with the ghostly reflection from the video.

Figure 3-10 Place the case and the figurine(s).

Step 3: Perfect View

Your video needs to have a dark background and a well-lit subject to work well. Play your video, and hold your phone over the center of the CD case as in Figure 3-11. Figure 3-12 shows you what the illusion looks like from the viewer window.

Get a friend to look through the viewing window to adjust the distance your phone is away from the front of the box until he or she can see the whole video. Be sure to think about whether you want to use the landscape

Figure 3-11 Positioning phone.

Figure 3-12 Viewer window.

or portrait position. If your partner can see other parts of the phone, that's okay for now. Once you get the right position, trace the phone around or measure how far in the phone is from the front of the box.

Before you cut, measure your screen size. Start by cutting a hole ½ inch smaller than your screen when it plays the video. This will allow you some wiggle room for adjustment. Resize the hole if necessary, and if you cut too much away, just tape or glue some black paper over the hole (Figure 3-13).

Now place the lid on the box, play the video on your phone, and position your phone facing down over the hole. Just as in the simple version of Pepper's Ghost, you will want to line the box with black paper to absorb light and prevent reflections (see Figure 3-13).

Figure 3-13 Illusion in perfect position.

Step 4: Backdrop

Make a backdrop and set for your illusion with construction paper, modeling clay, or Legos. Think about ways to cut and fold paper to make your backdrop. In this instance, we decided to make a meadow with a row of tall grass to hide some of the wiring for lighting.

Step 5: Add littleBits

You will also discover that because the box is sealed, you need to use the method described earlier in this project to illuminate the figurine in the box. For this task, there are a number of items you can use from small flashlights, to LED tea lights, to littleBits. For this project, we are going to use a long LED (o2) littleBits light to illuminate the boy figurine (Figure 3-14). We will begin the circuit with a power bit (p1) connected to a fork bit (w7). From the fork module, we will use an output wire to run to the long LED light; we bend the light over in the front corner of the box. You can use tape or cut a small slit in the corner to hold it in place.

Next, add two more output wires to the fork. These wires are going to power our lighting in the background set. Depending on the size of your box, you may need to add another output wire to extend the length, if needed. If you want to add a strobe-light effect, you can put a pulse bit (i16) before the wire or where the two wires meet. This way you can adjust the timing outside the box (Figure 3-15).

Figure 3-14 Long LED bit.

Figure 3-15   Pulse and wires.

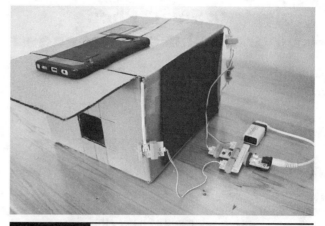

Figure 3-17   Close box and place phone.

Choose a LED bit (o1), bar-graph bit (09), or if you want color adjustment, choose an RGB LED bit (o3) to add to the end of the wires you have just run to the back of the box. You can hide the wiring and lights behind the props (Figure 3-16). Use the viewing window to make sure that they are hidden from the viewer.

Close the box, and place the phone with the video on top (Figure 3-17). If you can, try to put the video on repeat, if possible, and share your ghostly illusion! (Figure 3-18).

Figure 3-18   Ghostly dog illusion complete.

Challenges

- Create a background or stage for your figurine to appear on.

- Find a way to scare your viewer by allowing him or her to see nothing at first and by making your ghost suddenly appear later.

- Try adding a robotic figurine or a rotating platform using Legos or littleBits.

- How can you make your video interact with figurines and other backgrounds?

- Can you make the box relate to the illusion taking place inside it?

- How can video effects and lighting be used to give your video a spooky feel?

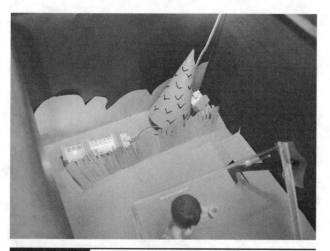

Figure 3-16   Hiding wires and lights.

Project 10: Smart Phone Hologram/Illusion Challenge

Cost: Free–$$$

Make time: 10–15 minutes

Supplies:

Materials	Description	Source
Cardstock	Cardstock for template	Recycling bin
Plastic	Transparencies are too thin, but you could use the plastic from a salad or fruit container	Recycling bin
Black paper	Black paper to help image appear	School supply store
Tape	Clear tape	—
Ruler	—	—
Scissors	—	—
Marker	Fine-tip permanent marker	—

Step 1: Create a Template

Use a ruler, pencil, and graph paper to create a template for the triangle. The bottom of the triangle will be 6 centimeters long. Make sure the center at 3 centimeters is on a perpendicular vertical line.

Draw a 3.5-centimeter line from the center at 3 centimeters that is perpendicular to the bottom. Instead of making a perfect triangle, we are going to slice the tip off. Measure 0.5 centimeter to the left and right of the 3.5-centimeter line, and place a dot at each mark.

Connect the dots to create a 1-centimeter line that is parallel to the bottom line. Connect the right side of the top line to the right side of the bottom line, and do the same for the left side to complete a triangle without a tip. Refer to Figure 3-19.

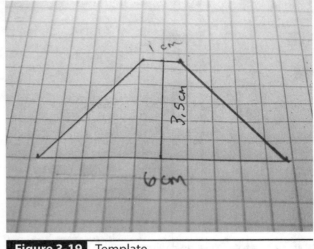

Figure 3-19 Template.

Step 2: Trace, Cut, and Tape

Clear plastic containers from baked goods, fruit containers, and toys work great for this project. Trim any curved edges from the recycled plastic containers so that you only have a flat sheet (Figure 3-20).

Place the plastic sheet over the template, and use a fine-tip marker with a ruler to trace the first triangle. If you can, try to plan it so that you can trace the next triangle using the side of the first one (Figure 3-21).

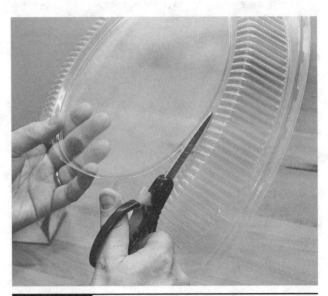

Figure 3-20 Trimming plastic sheets.

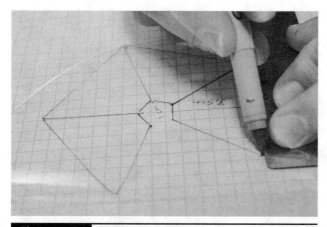

Figure 3-21 Tracing template.

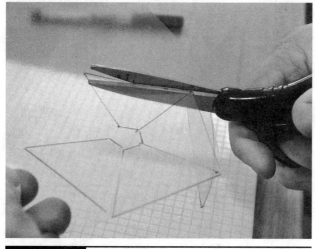

Figure 3-22 Cutting copies.

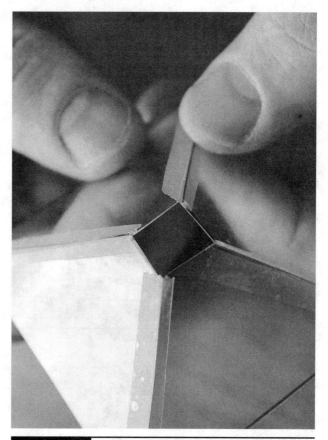

Figure 3-23 Taping triangles.

the pyramid will balance on your smart phone (Figure 3-23).

Step 4: Create a Dark Background

Place the pyramid on a piece of black construction paper. Trace around the pyramid on the paper, and then cut out the square. This will keep excess light and reflections out of your hologram.

Tape the square to the large opening of your plastic hologram, and the hologram is complete. You will notice that when the pyramid is assembled, the angles of all four side surfaces are roughly 45 degrees. Sound familiar? This smart phone "hologram," just like Pepper's Ghost, relies on reflection occurring on all four surfaces at a 45-degree angle. Figure 3-24 shows the completed hologram.

Cut all the triangles out, and lay a pair side by side (Figure 3.22).

Cut a piece of cellophane tape about 3.5 centimeters long, and cut it down the middle. Tape the triangle sides together. You should be able to get all the sides together except the last one.

Step 3: Fold and Tape Last Edge

You will notice when you pick up the triangles, they will fold easily one way. Fold your triangle in this easy direction along the sides to form a pyramid. Make sure that you match up the last seam evenly at the top and bottom so that

Figure 3-24   Complete "hologram."

Step 5: Create and Load Video

Find a hologram video on YouTube. These videos are easy to spot because they usually have four of the same images facing outward from the center. Place the small opening on the center of your smart phone, and enjoy your nineteenth-century parlor trick (Figure 3-25)! You can also put the large end of the pyramid on the table and balance your phone on the small end.

Challenge

- The true design challenge and maker fun actually lie in making your own hologram video or image. Think about how to turn the

Figure 3-25   Hologram illusion complete.

images to make them appear upright. Also consider the background and how it needs to contrast with the subject in the foreground!

Classroom tip: Pre-cut plastic strips 3.5 centimeters wide so that students can easily trace triangles and trim them in a short period of time. This should also help with accuracy and the levelness of the hologram.

Project 11: Smart Phone Projector

Did you know that you could turn your smart phone into a projector with just a magnifying glass and a cardboard box?

Cost: $–$$

Make time: 20–30 minutes

Supplies:

Materials	Description	Source
Box	Shoe box about 8 inches wide, 12 inches long, and 4 inches deep tall	Recycling
Paper	Black, gray, or dark blue paper	School supplies
Tape	Clear tape Blue/black masking tape or electrical tape	Office supply store
Scissors	—	Hardware store
Box cutter	—	Hardware store
Ruler	—	Office supply store
Lens	Wallet-sized Fresnel lens or small magnifying glass	Drugstore or Amazon

Step 1: Measure, Center, and Mark

If your box is less than 5 inches tall, find the horizontal and vertical center of the smaller end of the box. If your box is taller, make a mark at 5 inches high, and draw a line. This line will serve as the *top* of your box. Next, measure the size of lens (Figure 3-26). If you are using a credit-card-sized Fresnel lens, measure only the rectangle made up of concentric rings.

Calculate half the length and width of the lens. Measure that distance from the corresponding vertical and horizontal center, and make marks. Connect the lines to complete the rectangle. Check your math by placing the lens over the rectangle you just drew (Figure 3-27).

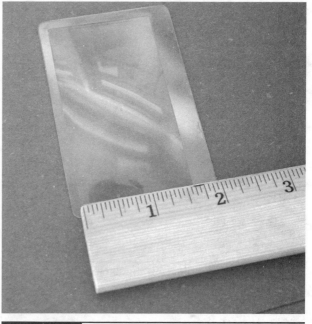

Figure 3-26 Measuring the Fresnel lens size.

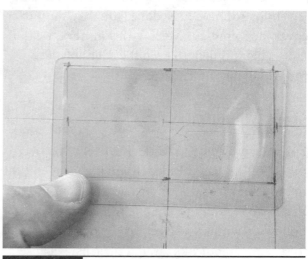

Figure 3-27 Drawing the Fresnel lens hole.

To use a round magnifying glass, begin by finding the diameter of the lens, or the length across the lens. Divide the diameter by 2 to find the radius. You will need to place a mark 1 inch away from the top and bottom of the vertical and horizontal center of the box. Align your lens, and trace around it (Figure 3-28).

Classroom tip: These instructions are for building a working example. We make our students figure out how to center the lens!

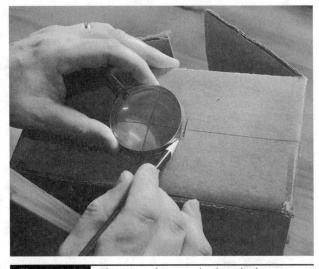

Figure 3-28   Planning the circular lens hole.

Figure 3-29   Curved lens in place.

Step 2: Cut it Out

Use a box cutter or craft blade to remove the lens hole. After you are done, you may want to cover the rough-cut edges with some electrical tape. This will help to keep the edges of the projection nice and sharp.

Step 3: Seal It

For your projector to work effectively, you need to seal off any light leaks. If you are using a round lens, place the lens in the hole, and use the handle to secure it to the box with some tape. Follow up with some electrical or masking tape to seal any light leaks around the lens. Because of the curved surface of the lens you may need to use smaller strips of tape (Figure 3-29).

If you are using a Fresnel lens, use the clear border around the Fresnel lens and some masking or electrical tape to seal it in place (Figure 3-30).

The inside of the box needs to be as dark as possible. If you have a light-colored or glossy box, you will want to use some dark construction paper to cover the inside of the box. Use a marker or black electrical tape to seal up any seams or spots you missed.

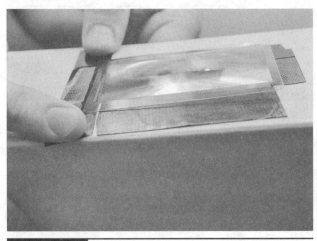

Figure 3-30   Fresnel lens in place.

Step 4: Get Focused

For this step, you will need a dark room, and you should turn the brightness on your phone all the way up. It is also handy to have some Play-Doh, a binder clip stand, or a pencil to help you hold your phone in place. Load a video or image full screen on your phone. Stand 3 to 4 feet away from the wall, and move your phone back and forth until the image is in focus. You will instantly note that the image is upside down and mirrored (Figure 3-31).

Take some time and experiment by adjusting your distance from the wall and the distance

Figure 3-32 Adjusting focus and projection size.

Figure 3-31 Projector image on wall.

your phone is from the lens. Try to achieve the best possible balance. Mark the optimal distance of the phone to the lens and the distance from the lens to the wall. Refer to Figure 3-32, where we are adjusting focus if you need help.

Classroom tip: It is tempting to have students lock the rotation on their phone before you complete this step; however, we think it is a great conversation starter. After students discover this phenomenon, ask them where they have seen or experienced this same effect in nature. It's a fantastic intro to the physics of light and how the human eye and our brains work together to allow us to see. It is also great to have both Fresnel lenses and convex magnifying lenses to make comparisons about image size and quality. If you are working with a group with several projectors, it is a great time to share the methods for achieving the optimal distance and focus. Let your students brainstorm some ideas about how to make a projector with adjustable projection size and focus.

Step 5: Flipping Out

Next, let's take care of the orientation of the image on your phone and correct the upside-down image. For iPhones, go to Settings → General → Accessibility → Assistive Touch → Device → and select Rotate Screen. For Android phones, you will have to download a rotation app that will allow you to lock rotation on landscape.

After you have your rotation correct, load a movie or a slideshow of your favorite photos, and turn off the lights! If you feel like mashing this project up with some littleBits and Makey Makey GO, head over to Chapter 12 to learn more!

Challenge

- Our rotation lock only solves the upside-down image. How could you make a projector that would also fix this problem without an app?

"Smart Stand"
Challenge

What is the cheapest and easiest solution for creating a stand for your smart phone? How can you use something like a simple paper clip to make a smart phone stand? Build your own smart phone stand with everyday office supplies or Legos, or even design and 3D print a stand. Can you create a stand to help you in your Pepper's Ghost illusion or smart phone projector?

   Make and take pictures of your challenge project. Tweet it to us @gravescolleen or @gravesdotaaron, or tag us on Instagram @makerteacherlibrarian and include our hashtag #bigmakerbook to share your awesome creations. We will host a gallery of your projects on our webpage.

Paper Circuits

HELP YOUR MAKERS learn about the circuits that power our world with these fun paper circuitry projects. Learn more about circuitry, and prep your maker muscles for more complicated electronics. Even lay the foundation for sewing circuits in Chapter 7. These beginner and advanced paper circuitry projects will flex your brain and help you to learn how to use circuitry in more complicated projects.

Project 12:	LED Origami
Projects 13–16:	Greeting Cards with DIY Switches
Project 17–18:	Pop-Up Circuitry Tricks for Pop-Up Books

Chapter 4 Challenge

Next-level paper circuit "pop-up paper circuitry book" challenge.

Project 12: LED Origami

Who doesn't love origami? Learn to fold a really simple bookmark, and add a functioning LED! You can even hack the technique in this project to light up any future origami projects. One of the great things about this project is that it can be adapted for almost any origami shape. You and your makers will learn more about different types of circuits in Projects 13 through 16.

Cost: $

Make time: 30 minutes

Supplies:

Materials	Description	Source
Origami paper	Assorted colors and styles of origami paper	Craft store
LEDs	Assorted 5-mm LEDs	SparkFun
Batteries	2032 coin cell batteries	SparkFun Amazon
Tape	Electrical or duct tape, clear tape	Craft or hardware store

Classroom tip: Bring in lots of different patterned and solid origami papers. Before beginning this project, you might decide to have origami design stations led by expert origami makers, where every table makes a different origami shape. Then you can help students to make a simple circuit with just one LED and a coin cell battery.

Step 1: Set Up Folds

Begin by folding the square origami paper in half diagonally. Turn the triangle so that the fold is at the top and the point of the triangle is facing you. Grab one of the corners along the fold and turn it down to meet the point at the bottom. Repeat this step of the opposite corner to create a diamond shape as in Figure 4-1.

Figure 4-1   Making a diamond.

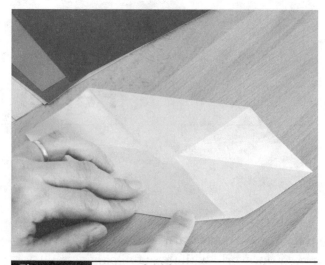

Figure 4-3   Repeat fold for top.

Step 2: Unfold and Refold

Now unfold the two folds you just did so that you have a triangle again (Figure 4-2). You may think it is crazy at first, but sometimes, in origami, the folds actually just serve as guidelines for other folds. Point the triangle toward you again with the fold at the top. Open the fold, and you have a square, but remember its creases we need. Take the bottom corner, pointing at you, and fold it over so that the tip is in the center of the paper. Line the edges up with the diagonal creases you made earlier. Repeat the step for the top corner now (Figure 4-3).

Step 3: Shape Shifting

Fold the paper down the center crease, tucking the triangles you just folded to the inside. You now have a trapezoid-shaped paper. Turn the trapezoid so that the short side is facing you, and turn down the corner of the long side so that it lines up with the center of the trapezoid. Repeat the step for the other side so that you have a diamond (Figure 4-4). Pick the paper up and turn it over so that you see the triangular fold on the back. Squeeze the shape slightly to reveal

Figure 4-2   Unfolded triangle with creases.

Figure 4-4   Trapezoid to diamond.

the pocket, and tuck the overhanging triangular shapes into the opening. This should leave you with a triangular shape that opens up when it is squeezed (Figures 4-5 and 4-6). We will need this fold later, but for now, undo the last folds so that you have a diamond again, and we can add our circuitry.

Figure 4-5   Diamond to triangle.

Figure 4-6   Triangle with opening.

Step 4: LED Eyes

We are going to make a very simple parallel circuit to power our LEDs. You can even test it before we install it. The long stem of the LED is the positive lead, and the short stem is the negative lead. To test your LED and battery, you can wedge a coin cell battery in between the corresponding leads on the LED, and it will light up (Figure 4-7). If it doesn't light up, try a different battery. For this project, you will need two LEDs to create the glowing eyes of our monster, bird, or whatever creature you want to design. Use a pin or thumbtack to poke two holes for each LED leg. The LED legs will go through the front of the paper and into the pocket we made, while the bulb will remain on the outside and function as supercool glowing eyes. You may want to line the LEDs up and mark them with a pencil. When you install them,

Figure 4-7   Testing LEDs.

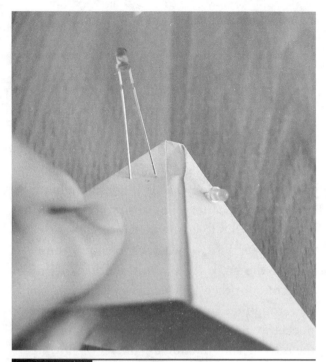

Figure 4-8 Push LEDs into position.

make sure that the two short sides (negative leads) are in the middle (Figure 4-8).

Step 5: Twist and Tape

Now that the LEDs are in position, we should have both negative ends in the middle. Use a pair of needle-nose pliers to twist the leads together (Figure 4-9). Push the twisted wires to one side,

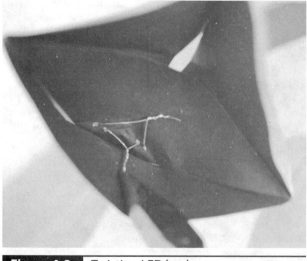

Figure 4-9 Twisting LED leads.

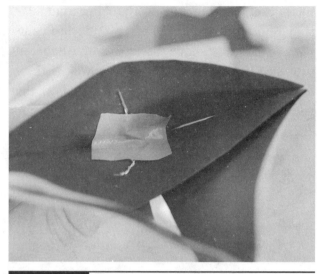

Figure 4-10 Using tape to insulate.

and then repeat the twisting process for the positive leads. Move the positive leads over to one side. Place a small piece of tape as close to the top corner as possible to insulate the positive and negative leads of the LEDs (Figure 4-10). This will keep them from touching before they connect to the battery.

Slide a battery in, and position it so that it touches the corresponding leads. Squeeze the sides of the triangle together to light your LEDs, as in Figure 4-11. If the LEDs do not come on, check the position of the battery, and make sure that you have the correct sides matched up. Use tape on the sides of the battery to hold it in

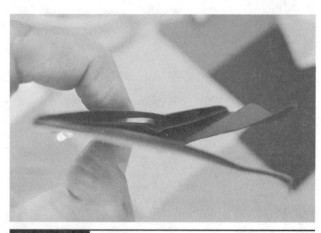

Figure 4-11 Squeeze test.

Figure 4-12 Taping battery.

place, making sure that the LED leads are still connecting to the battery (Figure 4-12).

Use paper to cut some eyes for your creation. You can cut a slit in the eye, and slide your LED through it (Figure 4-13) .We recommend that you spend some time trying different design combinations before you settle on just one. When you are ready, glue your eyes, ears, hair, or beak in place! This pocket origami animal makes a handy bookmark and rests neatly on the edge of a page (Figures 4-14 and 4-15).

Figure 4-13 Eyes.

Figure 4-14 LED bird.

Figure 4-15 Roosting on a book.

Classroom tips: Teach a few students how to make this beforehand, and then place your makers in small groups. Have your student leaders be the expert in each small group. Origami takes patience and careful listening, which is not likely to happen in larger groups. So keep your groups small with many experts guiding the learning.

Challenges

■ After students have mastered this project, fold any origami design you desire, and attempt to add your own circuitry.

■ How many LEDs can you add and your circuit still function?

- Could you add a tiny vibration motor and make your origami move?

Projects 13–16: Paper Circuit Greeting Cards

One of our favorite ways to teach paper circuits is by making our own do-it-yourself (DIY) greeting cards. Starting students off with templates and cute drawings lets your makers concentrate on building circuits and gives them confidence in their circuitry. Once they've understood and mastered circuits, give them cute drawings and have them design their own circuitry to light up the artwork (Figure 4-16). Lastly, let their newfound creative confidence

guide them into designing and creating their own artwork and their own circuitry!

Project 13: Simple Circuit Card

Current flows in a loop to complete a circuit. The electrons flow from the positive end of your battery, through the LED, and back to the battery's negative end. This closed path is called a *circuit*! Your project will only work if you use conductive tape (or other conductive material, like aluminum foil, conductive thread, etc.) because electrons flow through the copper tape to light up the LED.

If you wire up your LED the wrong way, you won't have a complete circuit, and your LED

Figure 4-16   Cards and Chibitronics stickers.

won't light up. For clarity, we've marked the "positive traces" and "negative traces" on the templates, but the negative "path" (normally called *routing* in electronics) isn't any different from the positive "path." Instead, think of it as a guide to help you lay your components out correctly and craft your own working circuits (Figure 4-17). Because paper circuits rely on batteries, they rely on a direct current (dc). Some components have a polarity (one side must be connected to a positive and the other connected to a negative), and some do not. For instance, LEDs depend on the polarity in your routing to function. If you hook them up incorrectly, they won't light up. Other components, such as an ohm resistor, will function no matter which way you hook them up because they do not depend on the polarity in your routing. However, it is important to note that there are some components that can be damaged or "fried" if you hook them up backwards. So always be aware of your routing and your components. We'll look at a few different types of circuits in this chapter and then combine these superpowers with coding to complete the Arduino projects in Chapter 5!

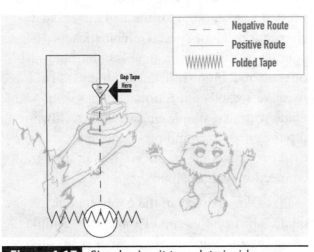

Figure 4-17 Simple circuit template inside.

Cost: $

Make time: 30 minutes

Supplies:

Materials	Description	Source
Template	Simple circuit template	Back of this book
Conductive tape	Copper tape (PRT-10561) or copper tape adhesive conductive (PRT -13827)	SparkFun
LED stickers	Chibitronics LED stickers: set of 30 or, for a makerspace, get a Circuit Stickers Classroom Pack	Chibitronics
Battery	2032 coin cell battery	SparkFun Amazon

Step 1: Make Copies

Make copies from the simple circuit template at the back of this book, and gather supplies.

Classroom tip: If you are teaching a class about paper circuits, make sure that you print your cards on bright and colorful paper. Also address some common issues with novices to paper circuitry. In most cases, only the topside of copper tape is conductive. It also needs to be applied in a smooth and even manner. Lastly, unless directed, you should use only one continuous piece of copper tape for circuit routing to ensure the best connections. We will point out clear tips as we craft through the next three projects, so let's make some great stuff!

Step 2: Create a Makeshift Battery Holder and Begin Circuitry

We are going to make a copper sandwich for your coin cell battery. Create a copper flap (as seen in Figure 4-18) that you can use as a switch for your simple circuit. As you stick the conductive tape down, press with your thumb to create a flat, even surface. When you get to the corner, press back on your tape and crease it; then turn the tape to fold the corner.

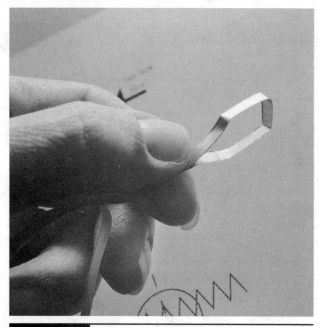

Figure 4-18   Create a copper flap.

Step 3: Mind the Gap

When you get to the spot for the LED, tear the tape. You need to leave a gap here so that the electrons can flow through the LED. If you were to lay the copper tape from one end of the battery to the next and lay your LED *on* the continuous tape, what do you think would happen? Your LED would not turn on because current is lazy! It prefers to take the path of least resistance, so the electrons would flow through the tape *under* your LED, and your LED would not come on. So tear that tape!

Step 4: Negative Traces

After the gap, lay down a new continuous piece of copper tape that leads back to your battery. This trace needs to go under the battery to complete the loop or circuit for the electricity to flow (Figure 4-19).

Fix tape as needed by pushing with your fingernail. Remember that we don't want our traces touching the wrong sides of our LED, so make sure that the gap is wide enough for that lazy electricity! We are almost ready for light!

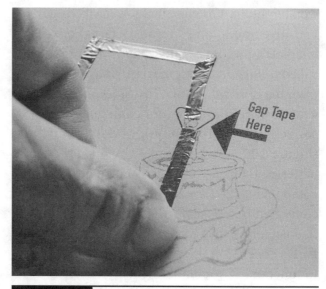

Figure 4-19   Mind the gap.

Step 5: Place LED

Minding polarity, place your LED sticker, and press firmly to the card where indicated on the template. The copper pad on the positive side of the LED needs to make contact with the positive trace and the negative trace needs to make contact with the copper pads on the negative side of your circuit sticker. Sometimes you'll notice that your LED only lights up when you press on the sticker. It might be that you haven't pressed hard enough for the LED to make a connection. These special circuit stickers do have conductive adhesive on the back, but they also offer a small bit of resistance. So press firmly to make a good connection, and let's see if you made your first simple circuit! You can always flip the paper over and press on the back of the card to ensure that the sticker is making a good connection with the tape.

Place your battery in the copper sandwich you made. The negative end of the battery should be on top of the negative trace (or path) on the right-hand side of your circuit. The copper flap from the positive trace should be over the top of your battery. The LED will only come on when you press the copper flap to the battery,

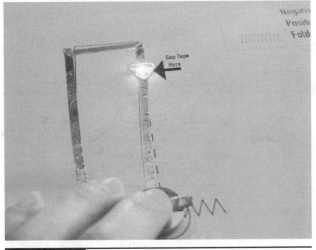

Figure 4-20 Test.

as shown in Figure 4-20. You can use a small binder clip to hold the battery in place and ensure a good connection. When finished, your simple circuit card should look like Figure 4-21.

Challenges

- Design your own light-up card. What would be a good illustration for a simple circuit?

- Use the same card design, but craft your circuit yourself. Can you make a smooth circle with rounded edges instead of a square?

- Can you add a story to your card? Or combine your light-up card with the origami project?

Classroom tip: Go over paper circuitry tips multiple times with makers. Ensure that they lay the copper tape down very smoothly, leave a gap for the LED (but not too wide!), and place the LED over the copper tape and not vice versa.

Figure 4-21 Final card.

Project 14: Flip the Switch Card

You might remember that we are huge DIY switch fans from our brush bot projects in Chapter 2. DIY switches in a paper circuit are even more fun because they can be part of the card design! In this project, you'll create your own switch to turn on a paper flashlight (Figure 4-22). Ready? Let's go!

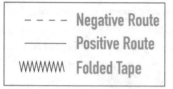

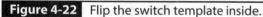

Folded Copper
Tape Flap Switch

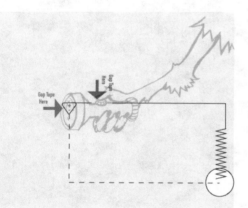

Figure 4-22 Flip the switch template inside.

Cost: $

Make time: 30 minutes

Supplies:

Materials	Description	Source
Template	Flip the switch circuit template	Back of this book
Conductive tape	Copper tape (PRT-10561) or copper tape adhesive conductive (PRT -13827)	SparkFun
Circuit stickers	Chibitronics LED stickers: set of 30 or, for a makerspace, get a Circuit Stickers Classroom Pack.	Chibitronics
Battery	2032 coin cell battery	SparkFun Amazon

Step 1: Copper Flap for Battery Switch

This copper sandwich trick that we learned from SparkFun is pretty amazing. Let's use it again starting at the battery side of the positive trace. On your template, this is marked as a zigzag. Create the flap the same way you did in the preceding project.

Step 2: Crease Tape

Remember to crease your tape when folding a corner. The easiest way to do this is to lay down tape up to where you want your corner and then pull it back a little. Then, using your thumbnail, put a crease in the tape. As you turn your tape at a 90-degree angle, use your thumb to put a nice fold in the tape, as in Figure 4-23. Smooth over this fold with your nail, and continue the tape trace until you get to the button on the drawing. Tear the tape because you do not want the switch to close until the DIY switch is pressed.

Figure 4-23 Fold corner.

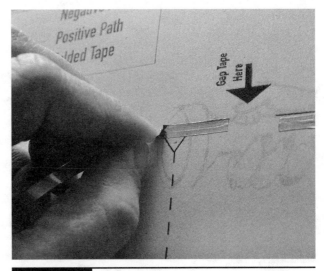

Figure 4-24 Leave gap.

Step 4: Negativity

You will now finish your circuit by creating a negative route with your copper tape. Starting under the negative side of your LED, smoothly place the copper tape. Push your tape back to make a crease for the fold, as in Figure 4-25, and fold your corner so that it looks like Figure 4-23. Following the path, finish your negative routing to end under where you will place your coin cell

Step 3: Mind the Gap

You are leaving a gap here as you did with the LED so that your user can turn on the paper circuit flashlight by pressing your switch and completing the circuit. With the tape gapped here, your circuit is open, and electrons cannot flow. However, once you have the card closed and you press on the drawing, you will complete the closed circuit, allowing electrons to flow and lighting up your flashlight.

Take a very small piece of tape, and start from the other end of the gap ending under the positive side of your LED, as in Figure 4-24. Leave a gap here as well so that the electrons will flow through your LED and light up your flashlight.

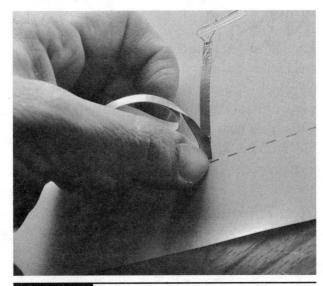

Figure 4-25 Crease.

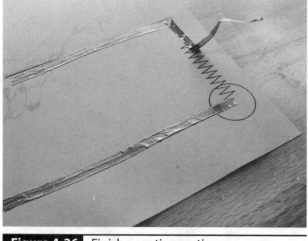

Figure 4-26 Finish negative routing.

battery, as in Figure 4-26. In Figure 4-26, we left the copper flap up so that you can see exactly how to sandwich your battery.

Step 5: DIY Switch

Using the copper flap trick, you will create a DIY switch for your paper flashlight. Fold the tape over about 1 inch, and then adhere to the card as designed in the template. We made two flaps just in case our drawing was off, as you can see in Figure 4-27.

Step 6: Light Up!

It's time to test your circuitry. Place your coin cell battery positive side up in the copper sandwich. You can use a binder clip to hold it in place or put cellophane tape over the sandwich. Close your card. It should only light up when you push on the button on your paper flashlight, as in Figure 4-28.

Step 7: Troubleshoot

Not lighting up? Check to make sure that the copper flap switch is aligned with your copper tape routing. Check to make sure that you have a gap under the LED and that your copper

Figure 4-27 Create copper switch.

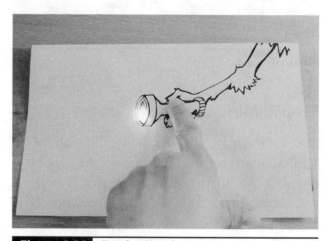

Figure 4-28 Finished card.

tape sandwich isn't accidentally shorting out the battery by touching both the cathode and anode of the battery (aka positive and negative terminals).

Always on? Make sure that you have a gap on your copper tracings that will only come on when you press the paper button.

Challenges

- What other drawing could you draw for this circuit?

- What other things with switches would make an intriguing card?

- Could you make your switch with a different method?

Project 15: Parallel Circuit Card

You can create series circuits with paper circuits and LEDs, but they require using multiple batteries. So, instead, we'll make a parallel circuit by placing our copper tape traces very close together and running LEDs across our parallel routing (Figure 4-29). In our simple circuit, the electrons flowed from our battery through the LED and back to the battery. In a parallel circuit, the electrons do a similar thing, but if we lay out our LEDs correctly, we can light up multiple LEDs, and the electrons will flow from the battery and through all our LEDs and back to the battery. Because of this, we can end our copper tracing at the last LED. You can continue the tape past the LED for aesthetic purposes, but it isn't necessary for our electron routing purposes.

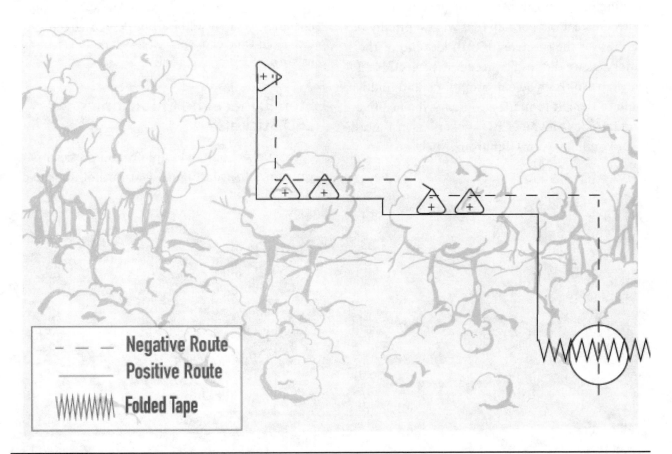

Figure 4-29 Parallel circuit template inside.

Cost: $

Make time: 30 minutes

Supplies:

Materials	Description	Source
Template	Parallel circuit template	Back of this book
Conductive tape	Copper tape (PRT-10561) or copper tape adhesive conductive (PRT -13827)	SparkFun
Circuit stickers	Chibitronics LED stickers: set of 30 or, for a makerspace, get a Circuit Stickers Classroom Pack	Chibitronics
Battery	2032 coin cell battery	SparkFun Amazon

Step 1: Copper Flap and Positive Route

Starting on the positive trace of the battery route, create a copper flap that will eventually be placed over the positive (or cathode) end of the battery, as we did in the preceding project. Fold your tape so that you can bring it up and fold it again to the left to make a route for the positive (anode) sides on your LED stickers, as in Figure 4-30. You'll lay your Chibitronic stickers on top of this tape, so make sure that your tape is smooth, as in Figure 4-30, and finish this route at the top end of your card (which will light up a hidden star!).

Step 2: Negative Route

Fold back your copper flap on your positive trace so that it doesn't interfere with your negative routing! Lay your tape flat, and follow the negative routing laid out on the template. Make sure that while you are following the tape, your LEDs will lay over the tape and connect both positive and negative sides on your LEDs in case your template is slightly off. End this route at the last LED.

Classroom tips: Make sure that the parallel copper tape routing is very close together but not touching. You want to ensure that both the positive and the negative sides on the circuit sticker will reach the designated copper tape route. Otherwise, you'll waste your copper tape and have a nonfunctioning circuit!

Step 3: Place the Chibitronics LED Stickers

Place stickers on the template, and press with your thumb, as in Figure 4-31 for both positive

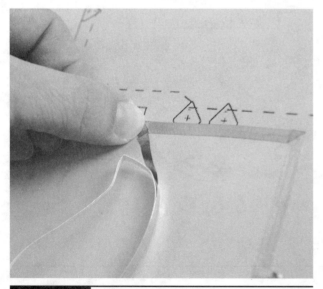

Figure 4-30   Route for positives.

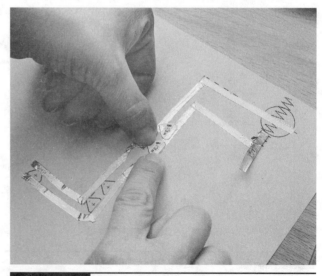

Figure 4-31   Place LEDs.

and negative traces, to make sure that your LEDs make a good connection with your copper tape wiring.

Step 4: Surprises!

One of our favorite things about paper circuitry crafts is that you can design hidden surprises for your user. Draw some hidden surprises for your card, as in Figure 4-32, making sure that they line up with where the light will shine through on the front side of the card.

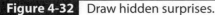
Figure 4-32 Draw hidden surprises.

Step 5: Test

Place your battery in the copper sandwich, and clip it with a binder clip, as in Figure 4-33. Did all lights light up? Hooray! No? Let's do some quick troubleshooting. Press down on all the traces of copper tape, and make sure that the tape is smooth and flat. Also press down on the LEDs to make sure that they are getting good connections. If they blink while you press on them, it means that they aren't getting a good

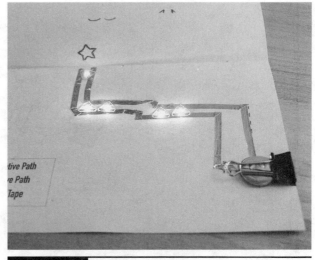
Figure 4-33 Test.

connection for the electrons to flow. You may need to flip your card over and press from the opposite side. Turn your card over and apply pressure to the routing and the LEDs. On the right side of the card, you can always use nonconductive clear tape over the LEDs to add more pressure, but just make sure that you don't get confused and use your conductive tape!

Make sure that all LEDs are facing in the correct direction with all positive sides on the positive route and all negative sides on the negative route. Double-check your battery placement as well.

Figure 4-34 shows the hidden bush monsters that appear when you light up your card and the twinkly nighttime star!

Challenges

- Can you make a parallel circuit in a circle?
- How about designing art with the copper tape routing? Can you make your parallel routing in the shape of a star?
- Could you add a DIY switch to your parallel circuit?
- Could you make AND/OR gates with your paper circuitry?

Figure 4-34 Final card.

Project 16: Disco Branched Circuit Card

You may have noticed on a light switch at home that sometimes there is only one switch for several lights. Other times there are several different switches on the same switch panel that turn on a different set of lights when they are flipped. We are going to use the same method to branch electricity and make a blinking disco light (Figure 4-35).

Cost: $

Make time: 30 minutes

Supplies:

Materials	Description	Source
Template	Disco branched circuit template	Back of this book
Conductive tape	Copper tape (PRT-10561) or copper tape adhesive conductive (PRT -13827)	SparkFun
Circuit stickers	Chibitronics LED stickers: set of 30 or, for a makerspace, get a Circuit Stickers Classroom Pack.	Chibitronics
Battery	2032 coin cell battery	SparkFun Amazon

Begin your copper tape routing as in Figure 4-37 for the positive side of your battery. At the corner, push your tape back to make a crease, and fold the tape as in the other paper circuit projects. End this routing with a copper flap that you will eventually place over the coin cell battery (Figure 4-38).

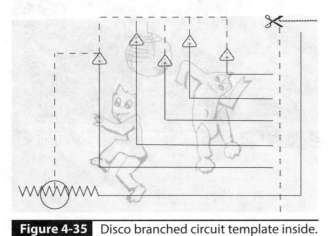

— — —	Negative
——	Positive
··· —··	Fold Paper Here
WWWW	Folded Tape

Figure 4-35 Disco branched circuit template inside.

Step 1: Prepare the Switch!

Following the template guide, cut the marked line so that you can fold your paper as in Figure 4-36. This is going to be the slide switch for our multiple LEDs.

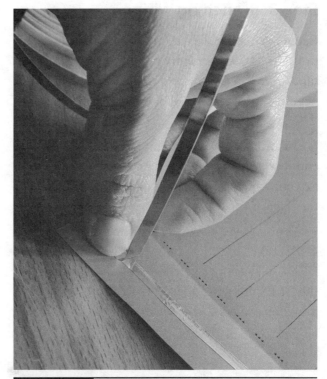

Figure 4-37 Positive switch.

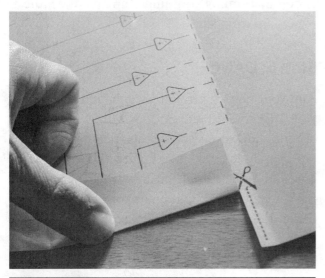

Figure 4-36 Cut and fold.

Figure 4-38 Copper flap.

Step 2: Negative Trace

Start at the negative side on the template under where you will place your battery. Place your copper tape along the template. Lay out one piece of tape that all negative branches will lead to.

Step 3: Branching Wiring

Begin branching wiring for the negative traces as in Figure 4-39. Remember to smooth the copper tape as you lay it out, ending just under the negative side from the LED circuit.

You will also need to make branches for the positive sides of your circuit stickers. These will end just under the flap from step 1 so that you can control your LEDs by sliding a finger across the flap.

As you make branches like those in Figure 4-40, make sure that you leave a gap for the LEDs so that the electricity will flow through the LED and not under it! Fold corners as in previous projects, and make sure that you are

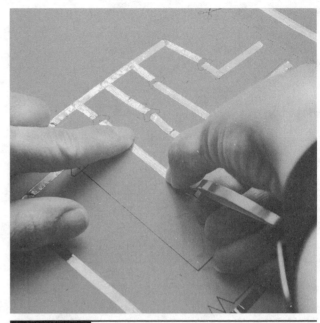

Figure 4-40   Mind the gap in bottom branching.

laying the copper tape down smoothly to create a smooth route for the electrons.

Step 4: Place LEDs

Place LEDs as pictured in the template. Push down with your thumb and fingernail to make sure that each side of the LED makes a good connection to the copper tape. Place your battery in the copper sandwich, and test your circuit by running your finger up and down the paper flap from step 1.

Step 5: DANCE!

Run your finger over your switch, and light up your LEDs. Close the card, and draw a silly electric slide, or just wow your friends with your blinking disco card, as in Figure 4-41. Figure 4-42 shows the completed card from the inside. Each LED should light up as your run your finger down your switch because each branch is making a connection with the positive trace on the flap as the pressure completes the circuit.

Figure 4-39   Begin negative trace.

Figure 4-41 Finished card.

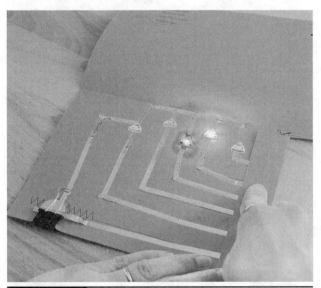

Figure 4-42 Inside of finished card.

Challenges

- How else could you branch your circuit?
- In what other ways could you use this type of switch?
- Can you design your own artwork for this circuitry?
- Chapter 12, the "makerspace mashup" chapter, adds another interactive element to this project, go check it out!

Classroom tip: This is a pretty complicated paper circuit and should only be done after makers have created and designed many of their own paper circuits based on the first three projects. Makers will need a good grasp of working with copper tape before attempting this project.

Projects 17 and 18: Pop-Up Paper Circuitry Tricks

Now that you know more about paper circuits, can you use them to make your own electronic pop-up books? Pop-up books rely on interesting interactive pull tabs. These projects will give you two inventive ideas on integrating circuitry into paper engineering and crafting your own pop-up book. We will leave the storyline and artwork up to you!

Project 17: Pull-Tab Switch

Cost: Free–$

Make time: 60–90 minutes

Supplies:

Materials	Description	Source
Paper	Assorted cardstock	Craft store
LEDs	Assorted 5-mm LEDs	SparkFun
Conductive tape	Copper tape (PRT-10561) or copper tape adhesive conductive (PRT -13827) or soft conductive tape from Makey Makey Booster Pack	SparkFun Joylabz
Battery	2032 coin cell batteries	SparkFun Amazon

Step 1: Create a Pull Tab

First, you need to create a pull tab for your page. Create an arrow that is about 5 inches long and 1 inch wide. The arrowhead base will be 2 inches long. Make a great arrow by measuring ½ inch on both sides of the shaft. Find the center of your arrow (about 1 inch) to make the point of the arrowhead. Draw it out with a ruler, and cut it with an X-Acto knife.

Step 2: Mark Slots and Aim Arrow

Use your arrow and a ruler to mark the slots where you will cut four slots on your paper for the arrow to slide into. Make sure that you use

your arrow as a guide for width. If you cut the slots too wide, you will have trouble with your circuits. However, you can always use a tiny piece of nonconductive tape on the inside to correct the width of a slot. Place the arrow on the paper so that the base of the point will be just at the edge of the paper, as in Figure 4-43. Mark your

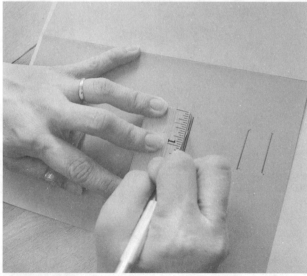

Figure 4-43   Marking slots on paper.

second slot around 1 inch away from the first, and then put your third slot only about ½ inch away from the second. The last slot will be 1 inch away from the third. The two narrow slots will function as a placeholder for the arrow and keep it close to the paper to aide in the connections for our electrons and to keep the arrow from wiggling. The wider-spaced slots will house our open circuits, which will be closed when the DIY switch passes over them. Use an X-Acto knife to slice the markings for the slots. The side on which you've drawn the slots is actually the backside of the paper, where you will house your circuitry. Flip your paper over, and push your arrow inside and through to the next slot (Figure 4-44), weave to the front side for the fourth, and finish by placing the arrow back through to the backside for the last slot, as in Figure 4-45. Mark the end of the arrow for future reference. You'll be adding a breaker to the arrow to keep it from going past this point in step 7.

Step 3: Design Circuitry

Using a pencil, mark out your circuit routing as in Figure 4-46. We are going to use a similar routing as we did in Project 14. However, this

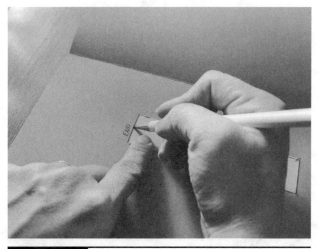

Figure 4-45   Mark the end.

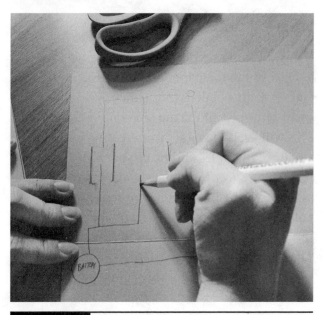

Figure 4-46   Designing circuitry.

time we will combine the branching technique from our Disco paper circuit (Project 16). We are including our circuit template in Figure 4-47, but it's time for you to start designing circuits on your own! You can do it! Make sure that you leave a gap for where you want to place your LED and one or two gaps in the large spots where your pull tab will slide across and function as a switch, allowing the circuit to close and the electrons to flow.

Figure 4-44   Tab inside card front.

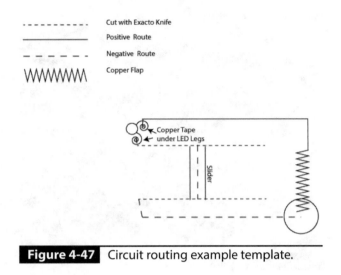

Figure 4-47 Circuit routing example template.

Step 4: Add LED

Using a sewing needle or small pin, poke two holes through the paper for your LED legs. You may want to stabilize your paper by adding a small piece of cellophane tape on the backside. This will keep the paper from tearing when you poke the holes and place your LED legs (known as a *lead*) through the paper. You'll want to mark your LED so that you remember the polarity. Here is a great tip from SparkFun: you can mark the bulb of your LED with a Sharpie! The long lead is the positive side, and the short leg is the negative lead. You should also see that the negative side is flat. Test your LED with your coin cell battery to be sure that both are working, and then mark your LED. Place the legs (leads) through from the front side of the paper. SparkFun has another great tip to make sure that your legs get a great connection: using needle-nose pliers, curl one lead and zigzag the other so that it has more places where it connects to the copper tape, ensuring electron flow (Figure 4-48).

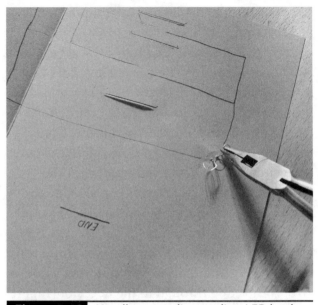

Figure 4-48 Needle-nose pliers curling LED leads.

Step 4: Lay Down Routing

Lay down your copper tape routing as you did in previous projects. You are becoming a pro! Make sure that you fold your copper tape corners neatly and keep your tape smooth. Are you ready to branch your circuit and use the pull tab to make a blinking switch? Just end one route of tape and start a new piece of copper tape routing to branch your circuit. Make sure to leave a gap so that the circuit is open until the tab is pulled, and then the closed circuit will allow the electrons to flow and light up your LED!

Make a copper flap for your battery because you will eventually use a binder clip to hold the battery in place. Continue your copper tape to your LED, and then lay the copper tape under the LED legs and tear the tape. Use a piece of cellophane tape to secure the LED legs to the top of the copper tape because the bottom of your copper tape may not be very conductive (Figure 4-49). Finish routing on the template, and now it's time to prepare your DIY switch!

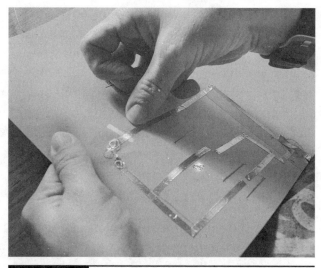

Figure 4-49　Adding the LED.

Step 5: Pull-Tab Switch

Place your arrow tab on your card so that you can mark where you need to create your copper switch, as in Figure 4-50. You will need to add some depth to the switch portion of your arrow to ensure that your copper switch will connect with your circuitry. Cut two small squares for the

spot you measured, and use a piece of double-stick tape to adhere the squares to your arrow, as in Figure 4-51. Cover this entire square with copper tape. I covered it horizontally and then used a piece of conductive tape on each end to cover the ends of the tape so that they wouldn't snag on the card when the tab is pulled (Figure 4-52). It also makes the switch look neater.

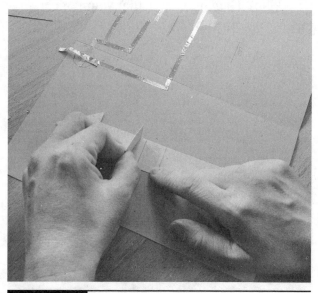

Figure 4-51　Double-stick tape.

Figure 4-50　Measure and mark for the switch.

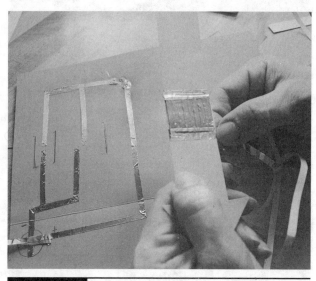

Figure 4-52　Adding copper to the pull tab.

Step 6: Battery and Test

Place your battery in your copper sandwich, and make sure that you align the positive end with the positive lead on the LED. Use a binder clip to hold the battery in place.

Test your switch by placing it in the spot where you want it to connect the circuit, as in Figure 4-53. Does the LED light up? Great work! No? Then make sure that your copper tape is smooth and one continuous strip (except for where you branched it, of course). Also make sure that the branched circuit is coming off the main circuit and applied smoothly. Does your copper switch connect your broken circuit? If not, you need to adjust the placement of your copper switch! If it still isn't working, make sure that your LED is getting a good connection with the copper tape.

Step 7: Finish Pull Tab

Put the arrow back through the slots as you did in step 2. If it's hard to get through, you may need to use your X-Acto knife to widen the slot. However, don't cut too deeply; your switch depends on the card holding the arrow close! Once you get it back in the card, you'll want to test it to make sure that the circuit works! When you pull your arrow, it should light up your LED. If your LED doesn't light up, try pressing with your fingers to see if it just isn't getting close enough to complete the circuit. If that works, use a piece of cellophane tape on the slot on the circuit side of the page. This is a quick fix if you cut too deep in any of your paper engineering projects! It can also help insulate circuits if you need it! Once your circuitry is working, you'll want to add a brake to your pull tab. Put the arrow where you want it to stop. Cut a small ½-inch strip from the same paper you cut your arrow from. On the inside of the card, tape the strip as in Figure 4-54, taking care not to tape the arrow to the card because then your pull tab won't pull! Now you can pull your arrow to this spot, and it will stop! This will keep users from accidentally pulling the arrow out of your book. Decorate with your own hand-drawn art (we have an example in Figure 4-55), and write a story to begin your own electronic pop-up book!

Figure 4-53   Testing the circuit.

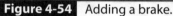
Figure 4-54 Adding a brake.

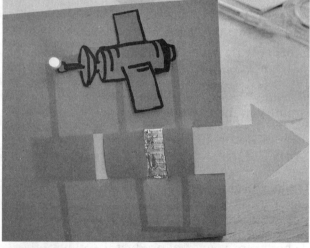

Figure 4-55 Decorate with art.

Project 18: Tiered Pull-Tab Trick

This is a great pop-up trick if you want to create an interactive character for your book. You'll create a slider and then attach it to a page in your pop-up book.

Cost: Free–$

Make time: 60–90 minutes

Supplies:

Materials	Description	Source
Paper	Assorted cardstock	Craft store
LEDs	Assorted 5-mm LEDs	SparkFun
Conductive tape	Copper tape (PRT-10561) or copper tape adhesive conductive (PRT -13827) or soft conductive tape from Makey Makey Booster Pack	SparkFun JoyLabz
Battery	2032 coin cell batteries	SparkFun Amazon

Step 1: Cut Papers

First, you need to cut your papers. You'll want to cut your base page about 1 inch larger than your second-tiered page. For example, we cut one rectangle 5 inches by 10 inches and the second paper 4 inches by 9 inches. After you cut them, fold them in equal thirds (Figure 4-56). You can use a ruler to maintain accuracy. Once you fold the papers, you'll need to mark where you need to cut your slots for your sliding mechanism. Mark about 1 inch down and 1 inch in from the left edge, as in Figure 4-57. Mark your other slot 1 inch in from the inside fold. Using your base page, mark where the slots of your slider will be housed on your smaller paper. Your slider will be on the far right side of the paper. Using an X-Acto knife, cut the slots for

Figure 4-56 Cut papers and fold in thirds.

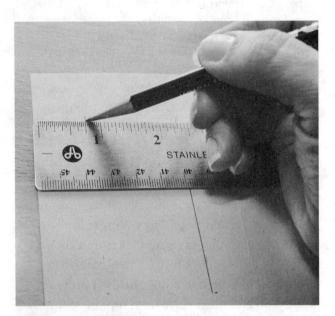

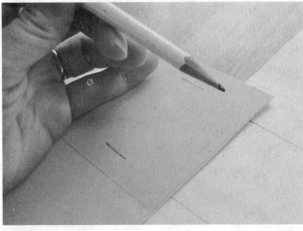

Figure 4-57 Mark slots.

Figure 4-58 Cut slots.

the sliding mechanism, as in Figure 4-58. Cut your slider out of a contrasting color. It should be about two-thirds the width of your top-layer paper (Figure 4-59). Use your slider to measure and cut the slots on the top-layer paper. Fold it in thirds so that you can use it for designing your circuitry in the next step.

Step 2: Draw a Simple Circuit

Draw your own simple circuit across the slider switch at the top of the slide, as in Figure 4-60. You can poke the LED through, but it would be better for your final product if you have the LED leads taped to the inside and the top of the LED peeking out of the top of the tier.

Step 3: Flap to Slide

Make a copper flap as in previous projects, and place tape up to the slider switch on the left, as in Figure 4-61. Wrap the copper tape around to the backside of the page to ensure a connection when the slider is in place.

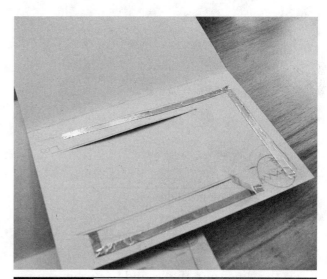

Figure 4-60 Draw a simple circuit with slider.

Figure 4-59 Cut slider and slider slots.

Figure 4-61 Copper flap and route to LED/slider.

Step 4: LED to Battery

With a new piece of tape, lay down copper tape from the LED leg to the battery. Place the tape under where the battery will be so that you can place the battery over this tape and under the copper flap from step 10. Place a small piece of tape from the other LED leg to the inside of the slot by the top of the right side of the slider.

Step 5: Electrons Slide

Prepare your sliding mechanism by putting copper tape over the middle third of the slide, and poke the ends of the slide through the slots on the base page, as in Figure 4-62.

Figure 4-62 Prepare slider.

Figure 4-63 Insert slider.

Step 6: Slide to Second

Insert the slider into the second tier, as in Figure 4-63, and tape the legs to the opposite side of the page. Use double-sided sticky tape, and do not get any tape outside the slide because you want this whole tab to move up and down.

Step 7: Attach LED

Curve the leg of your LED or make it zigzagged as in Figure 4-64 so that your LED will cover more surface on the copper tape. Once you get the legs in place over the copper tape, use regular tape to hold the legs down.

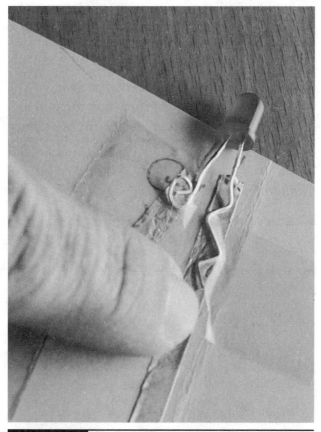

Figure 4-64 Prepare the LED.

Figure 4-65 Insert battery and test.

Step 8: Insert Battery and Test

Put your battery in, pull your second tier up, and watch your LED light up as in Figure 4-65. Does your circuit work? If not, you may need to add copper tape to the inside of the slot area to make sure that the electrons are able to flow across the slide. Also make sure that the LED legs are making a good connection with the copper tape.

Step 9: Hide Circuits

Almost done! Your circuit should be on the left side of the bottom tier. Fold the circuit toward the inside, as in Figure 4-66. Fold it all the way over, and place a piece of double-sided tape to hold it together. Ensure that you do not put tape over the second tier, or your pull tab will not function! Now that you have a working circuit as in Figure 4-67, it's time to transform this funky paper engineering into a character for your book!

Figure 4-66 Close pull tab and tape.

Figure 4-67 Transform into character and insert into story.

Step 10: Finishing Touches

Decorate the front of your character according to your storyline. We stuck with monsters and aliens and made a funky one-eyed monster whose hair accessory lights up when you pull on his hair! Insert your tiered pull-tab monster into your story! What will it say?

Challenges

- Can you nest another tab with a functioning switch?
- Can you light up more than one LED?
- What other type of pop-up switches can you make?

> ## "Pop-Up Paper Circuitry Book" Challenge
>
> Can you create a pop-up book now that you know so many paper circuit tricks? What other ways can you integrate circuits with pop-ups? Craft a storyline, and light up your book with LEDs! How can you make circuitry be an integral character in your story? Get tinkering and see what you can create. Once you learn to use Scratch and Makey Makey, you may decide to mash all these maker tools together to create one amazing project!
>
> Take pictures of your finished pop-up book, and tweet it to us @gravescolleen or @gravesdotaaron or tag us on Instagram and include our hashtag #bigmakerbook to share your awesome creations. We will host a gallery of your projects on our webpage.

Coding

IN THIS CHAPTER we have a couple of beginner projects with Scratch to get you started with computational thinking. In Project 21, we'll take that knowledge to the next level by looking at graphical programming with Ardublock and translating that into real lines of the Arduino coding language.

> Project 19: Getting to Know Scratch
>
> Project 20: Moving Around with Scratch Maze
>
> Project 21: Arduino littleBits Project Using Ardublock

Chapter 5 Challenge

Take coding further with a multi-level or Arduino Programming challenge.

Project 19: Getting to Know Scratch

Scratch is an amazing free resource from the MIT Media Labs. It's designed to get beginning coders confident in creating their own games, animations, and interactive stories.

Cost: Free

Make time: 30–60 minutes

Supplies:

Materials	Description	Source
Computer	Computer with Internet access	—
Scratch	Scratch account	scratch.mit.edu
(Optional)	(Optional) Raspberry Pi computer	SparkFun (DEV-13297)

Classroom tip: If you have young makers who have no experience with coding and programming, try code.org's Hour of Code before starting with Scratch. This will give them some basics in computational thinking but also strip down to basic programming and be a little less overwhelming. However, we believe that high school students with no programming experience can tackle Scratch with ease.

Step 1: Create Account

Create an account at https://scratch.mit.edu/. Login and click "Create" to get started on your first project. Name it "Dodgeball Game" or something similar so that you can find it again in the "My Stuff" tab.

Step 2: Get to Know Scratch, Get to Know All about Scratch

Let's take a moment to get familiar with your work area. On the left of your screen is the "Stage" or game play area, where you can see your game in action. Once you have your game fully created, you can click full screen mode to allow your game to take over your whole screen. Underneath the "Stage" is your "Sprite" area. In the middle of your screen is the "Blocks" tab/"Costumes" tab/"Sounds" tab. It defaults to the "Blocks" tab because this plethora of palettes houses all the Scripts (or building blocks) you will use to create your game or story. You can tell from first glance that many are self-explanatory. If you want to program sound effects for your character, where would you look first? To the right of the Scripts is your workspace, or "Scripts Area." To make a game or a program, you will drag blocks from different block palettes and snap them together in this workspace. Under the "Scripts Area" is your "Backpack." You can store scripts of code, also known as *programs*, in your backpack to use from one sprite/project to another sprite/project (Figure 5-1).

Step 3: Pick a Sprite

Sprites are the characters in your game. You can pick a sprite, upload a sprite, or even draw your own sprite. Sprites can also be used to control game play even if the sprite doesn't act like an active character in the game. In the next project, you'll create sprites that will start and stop your game but do not move around the screen. For now, look around in the sprite libraries and pick a sprite. To access the sprite libraries, look in the lower-right corner under the "Stage" window and the x/y coordinates. Take a look at the sprite list in Figure 5-2. The first icon is how you access the existing sprite libraries in Scratch. The pen icon here will allow you to draw your own sprite, the folder will allow you to upload a sprite, and if you are feeling funky, you can use the camera here to take a picture of yourself and be a character in your own game! (That's so meta.)

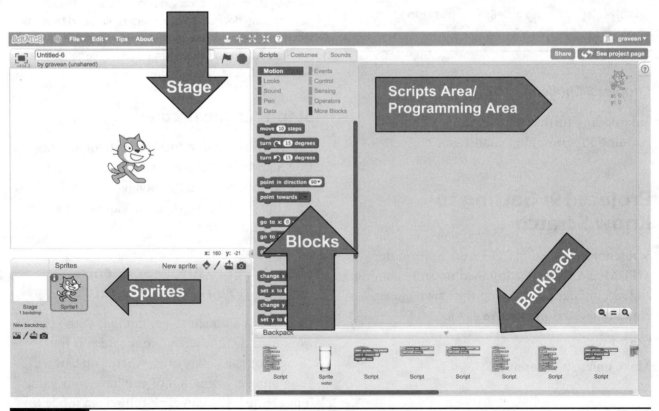

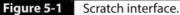
Figure 5-1   Scratch interface.

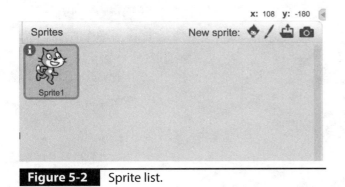

Figure 5-2   Sprite list.

Once you select a sprite, you can click on the "Costumes" tab to personalize your sprite (Figure 5-3). Play around here for awhile and discover what each of the tools allows you to do with your sprite. You can see that I'm changing the color of my sprite with the "Paint Bucket" tool. Can you make your sprite a patchwork of colors? From this menu, you can give your sprite many different "costumes," but for now, let's just

personalize your sprite and move on to finding a backdrop for our dodgeball game.

Step 4: Dress the Stage

Under the stage, on the left side of your sprites, you can upload a new backdrop. If you click on "Stage" while you are in the "Costume" tab of your sprite, you'll automatically be in drawing mode for your stage. Look at the close-up in Figure 5-4. You can click the first image icon in this menu to access the Scratch backdrop libraries. The pen tool allows you to draw your background (which we will be using in the maze project), and the folder allows you to upload your own backdrop. We decided to mess around and try uploading a rejected drawing from Chapter 4. You can adjust the size of the photo in the drawing area by selecting a portion of it and dragging it to be larger. This is a quick way

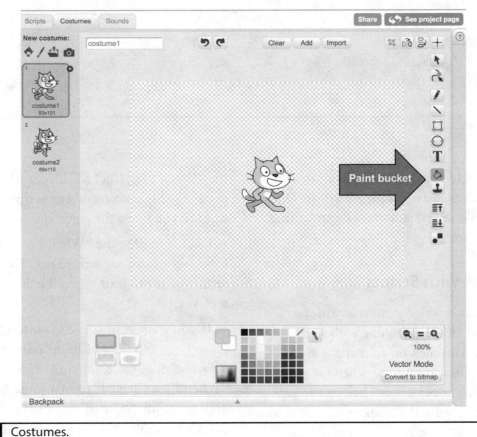

Figure 5-3   Costumes.

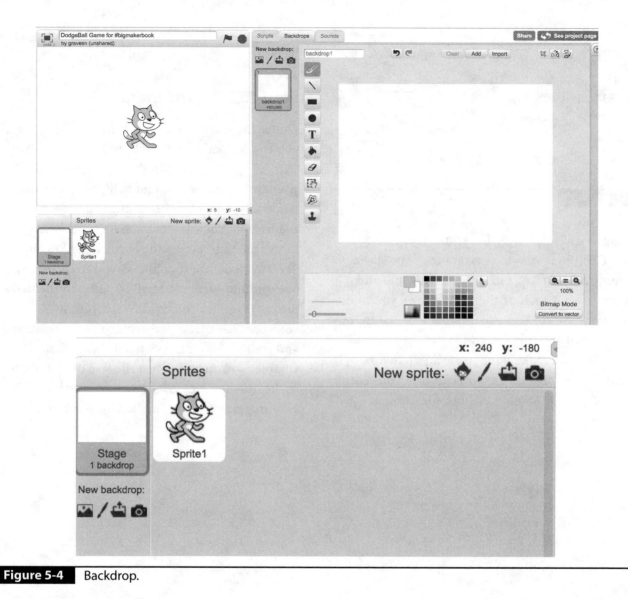

Figure 5-4   Backdrop.

to add a picture to your stage, but you can also easily scan and upload a unique backdrop. Once your backdrop is ready, let's move on to writing your first program!

Step 5: Work with Scripts

We need to select a script event that will start our program when the green flag is clicked. Make sure that you click on your sprite and place it where you want it to be when the game is started. In the "Blocks" tab, click on "Events" (Figure 5-5). These are the blocks you'll need to program your sprites movement. It also houses

the "When Flag Clicked" block, which you will need at the beginning of any scripts that you want to run when the game starts.

Click and drag the "When Flag Clicked" block/script to your script area. This script will restart your program when the flag is clicked by a player.

If you want your sprite to start where you've placed it, go to the "Motion" palette (Figure 5-6), and drag the command "Go to x: y:" to "When Flag Clicked" until a white highlight appears on the "Event" script, as in Figure 5-7. Connecting these together makes a small

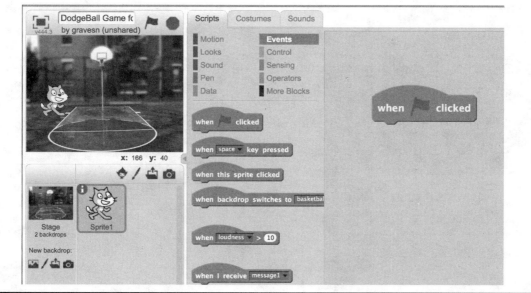

Figure 5-5 "Events" palette.

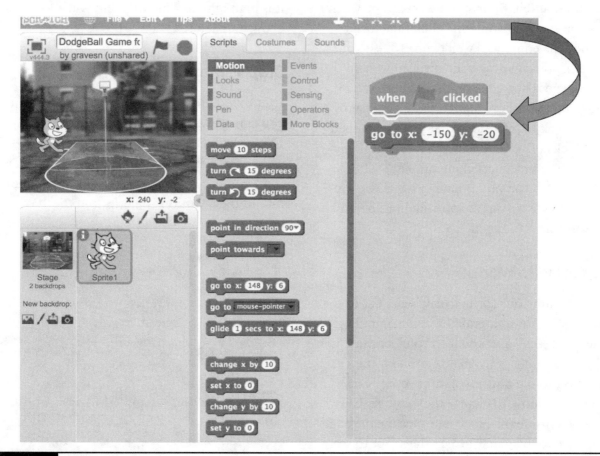

Figure 5-6 "Motion" palette.

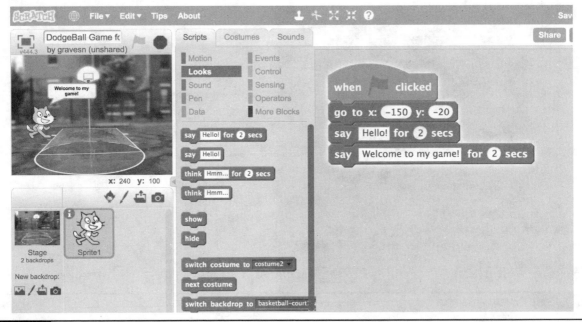

Figure 5-7   "Looks" palette in action.

program that tells your sprite where to be when the game starts. If you don't click the blocks together, the command won't run. To test your program, you can drag your sprite anywhere on the stage. Then click on the scripts to test it. Does your sprite hop back to its original spot? If not, check to make sure that you've snapped your blocks of code together. You can test any of your programs by clicking on them. This is a great way to test programming while you are working to see if your scripts function the way you want them to function.

Step 6: Hello World!

Let's program your sprite to say hello. Go to the "Looks" palette, and grab the command "Say Hello for 2 secs," and add it to your scripts. Click and drag another "Say Hello for 2 secs" block, and add it to the bottom of your script. This time we are going to change the message. Click in the box inside the block and type the message, "Welcome to my game!" Click the script to see your first "Hello World!" program in action! You'll notice when your script is running that

it is highlighted. This will help you a lot with debugging later!

Step 7: Move Your Sprite

It's time to program your arrow keys so that you can move your sprite. Drag a "When Space Clicked" block from the "Events" palette to your work area. Click the arrow inside the block to change "space" to the "up arrow," as in Figure 5-8. In the "Motion" palette, find the "Point

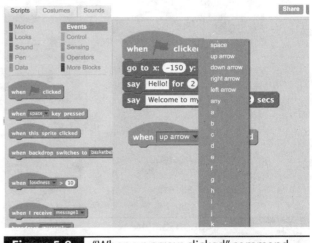

Figure 5-8   "When up arrow clicked" command.

in direction" block, and drag it to your "When up arrow clicked." Now your sprite will point up before moving up. Drag a "Move 10 steps" block to your script, as in Figure 5-9. If you don't want your sprite to point up when it moves up, you can change this setting by clicking the "i" on your sprite in the sprite list. The window will change to look like Figure 5-10. Change the rotation style to fit what you would like to happen in your game, and we'll get busy programming the other arrow keys in the next step.

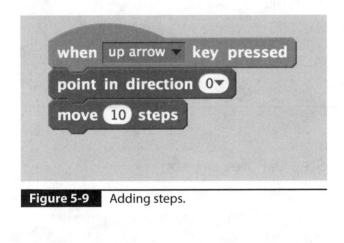

Figure 5-9 Adding steps.

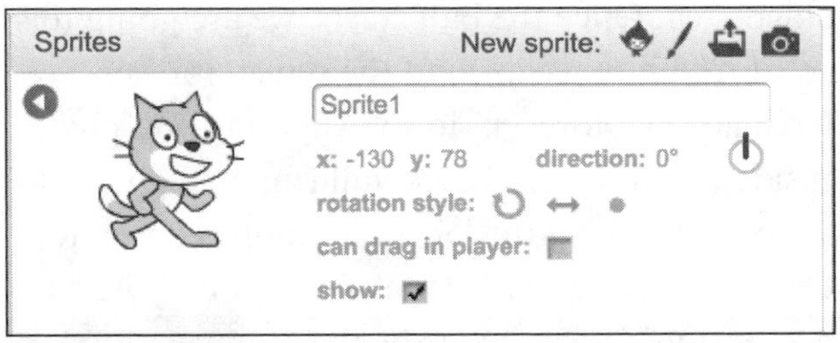

Figure 5-10 Sprite info.

Classroom tip: There are other ways to program your arrow keys. If your makers want to try to program their arrow keys in a way that is different from this project, let them! We'll show you another way to program arrow keys in the next project. Just remember, one of the best things about the maker movement is that there is no one right answer! It doesn't matter how you solve the problem; all that matters is the meaning your students make as they travel toward the solution!

Step 8: Duplicate Scripts

You can follow the same previous steps to program all of the keys, or you can duplicate your script by right-clicking on the script you want to copy and clicking "Duplicate." Then all you need to do is change your script to say "When the down arrow clicked" and change the x/y coordinates accordingly. Program all keys, and your scripts work area should now look like Figure 5-11.

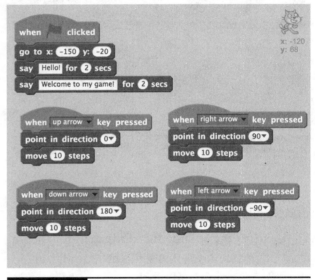

Figure 5-11 Scripts in motion.

Step 8: The Dodge Ball

Now that our sprite can run, let's give it something to run from! Click on "New sprite," and pick a ball sprite. If you want to change the color, click on costumes and adjust your sprite until your ball looks the way you want it to.

You may have noticed that all your scripts for your first sprite are gone! Not to worry, each sprite gets its own script work area (Figure 5-12). If you click on your first sprite, you'll see all your previous work. Click back on your ball so that you can program it to move. You are not going to program the arrow keys as we did with Sprite 1. Instead, we want our ball to bounce in

Figure 5-12 New sprite work area.

random directions as if we were really playing dodgeball.

We are going to hide this dodgeball for the start of game play, so let's go ahead and drag the "Show" command block from the "Looks" palette. We also want the ball to start in the upper-right corner, so drag your ball to where you want it to start, and then grab a "Go to x: y:" block from "Motions." Now drag a "Point in direction" block to your script. To make our ball move in random directions, you'll have to grab the "Pick random 1 to 10" block from the operators. Change the numbers inside the operation to read "4 to 140" as in Figure 5-13.

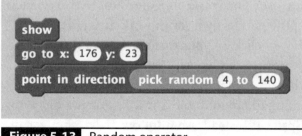

Figure 5-13 Random operator.

Step 9: Forever Loop

If you click your script to run it, the ball isn't doing much. We need to program your ball to move, but we also need to keep it inside the stage area. Once the ball starts, we are going to want it to bounce around the screen forever. So we'll need a "forever" loop. Anything you place inside the "forever" block will run, you guessed it, forever! Using the "forever" loop is a way to create a never-ending code! Therefore, if we put a "Motion" block inside of the "forever" loop, our ball will move around the stage ... forever! However, because there is nothing stopping it, it will also leave our stage area. Click the script to try it. Let's debug by adding a block to make the ball stay within the stage walls. Go to the "Control" palette and drag a "forever" block to your current script. Now add a "Move 10 steps" block from the "Motion" palette. You'll notice that with the script like this, the ball will bounce forever *and* be continually off screen. To keep the ball on the stage, you'll need to add a conditional statement. Inside the "Motion" palette, there is a very important block called "If on edge, bounce." This block will keep the ball inside the stage by telling the ball to bounce off the edge if it hits any of the four sides of the stage. Because it is located inside the "forever" loop, the ball will move 10 steps forever *unless* it hits the edge of the stage; *then* it will bounce back toward the inside of the stage! Whew! That was easy! See Figure 5-14 for the full script.

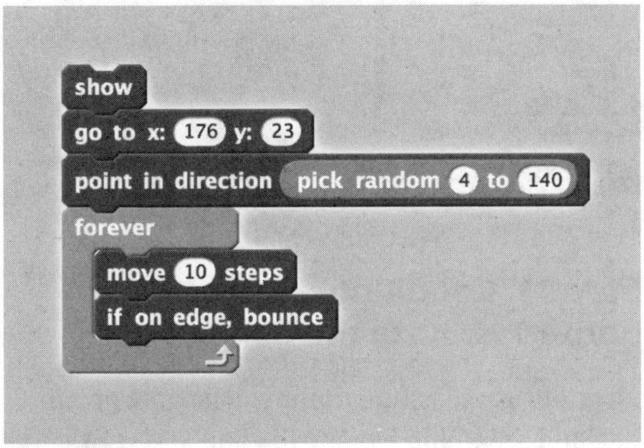

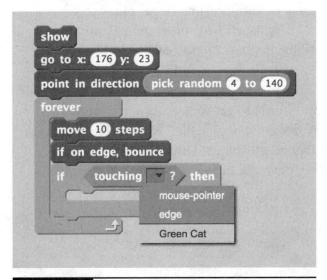

Figure 5-14 "Forever" loop.

Step 10: Conditional Statements: IF/THEN

Once you get your ball bouncing, you'll notice that it bounces right through our Sprite 1. We want the cat to know it is out of the game if the ball hits it. We also want the user to realize something has happened, so we'll need to add something to alert the player. We are going to add an "if/then" statement and a sound effect to make a pop sound when the ball hits Sprite 1. The "if/then" block is located inside the control statements. It is one of the most important pieces of code you will learn! This block is based on the conditional geometry statement that "if" this happens, "then" this should happen. We want to tell the game: if the ball touches Sprite 1, then play a sound effect. To do this, we will place the "if/then" block inside the "forever"

loop so that the ball will always be rolling but also be searching for the cat. We need to drag a "touching?" from the "Sensing" palette and place it inside the "if" block as in Figure 5-15. Make sure to click the drop-down arrow and click on the Sprite 1 character. For my script, it reads, "If touching green cat then." To add a sound effect, grab "play pop" from the "Sound" palette and nestle it inside the "if" block as in Figure 5-16. The ball will now play a pop sound when it hits Sprite 1!

Figure 5-15 Sensing.

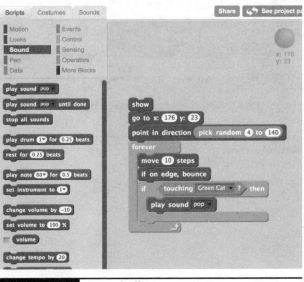

Figure 5-16 Sound effect.

Step 11: Get the Ball to Start/ Broadcast Messages

One of the coolest things about programming (and maybe also the most frustrating …) is that the program you create runs from the top down. Because of this, we do not want our ball to start before the user even understands the game. Instead, we will let our Sprite 1 tell the player how to play and then add a "Broadcast" script at the bottom of our Sprite 1 "Hello" program that will get our dodgeball to start running! Drag a "Broadcast message" block to your "When flag clicked" program on your Sprite 1. Name it "Start Dodgeball," as in Figure 5-17. We will also have to add a "Receive" script to our ball sprite to tell it to start. So click on your ball sprite to get back to its Script work area. You should now se the "When I receive Start Dodgeball" block! Drag it to the top of the script you've been working on, as in Figure 5-18.

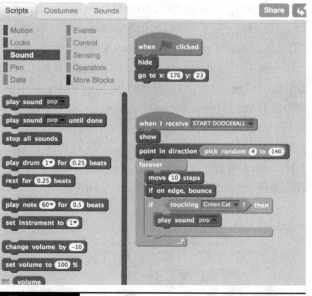

Figure 5-18 Receive message.

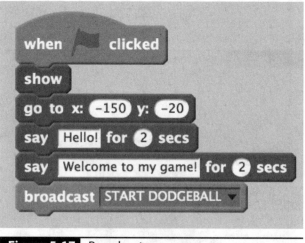

Figure 5-17 Broadcast message.

Step 11: Conditional Statements: Sprite 1 Is Outta the Game!

Now our game is functioning! Click the green flag and see if you can move your Sprite 1 to dodge the ball. You'll notice that the ball will hit Sprite 1 and make multiple popping sounds. In a normal game, what would happen? How can we improve our game play? Let's make Sprite 1 (or the "green cat" for my game) disappear when the ball hits it. We'll need another conditional "if/ then" statement. Drag the "if/then" block from "Control" blocks, and click it to our broadcast message. You'll need the "Sensing" block in your "if" statement. We want to tell the game that if Sprite 1 is touching the ball, then hide it. What do we need to put inside our "if/then" statement to make this occur? Look inside the "Looks" blocks for "hide," and add it inside the "if/then" block. Your script for Sprite 1 should now look like Figure 5-19. Play your game. What happens? Why do you think this is? How can you fix it? Try adding the "if/then" script to all the arrow keys as in Figure 5-20. Now what happens?

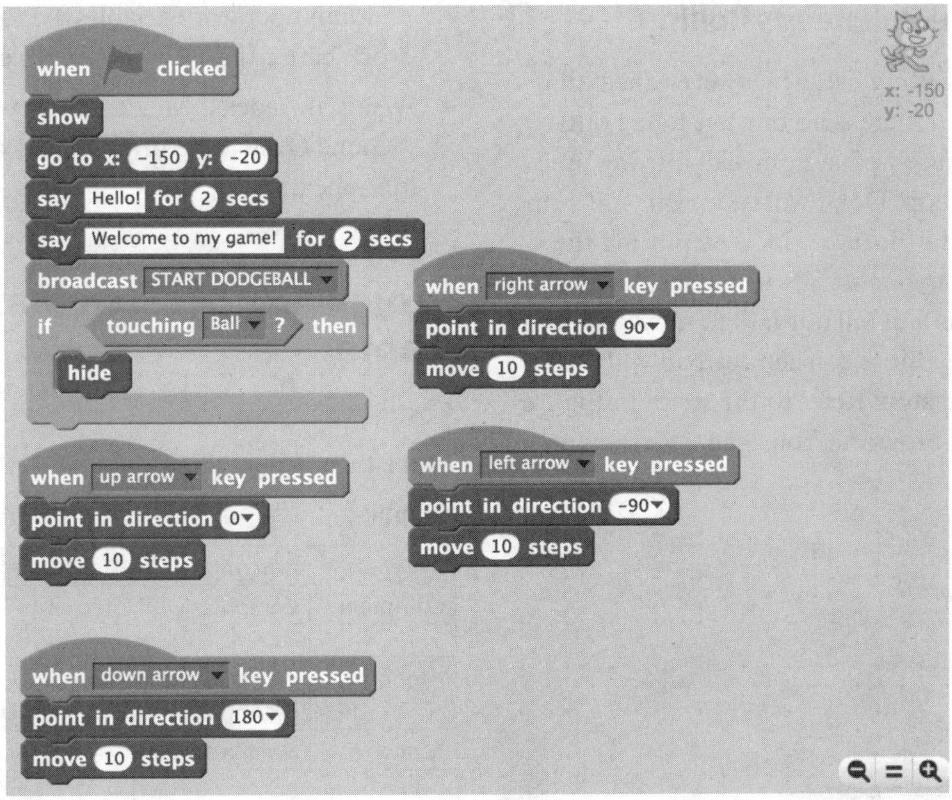

Figure 5-19   Make Sprite 1 disappear! You're outta the game!

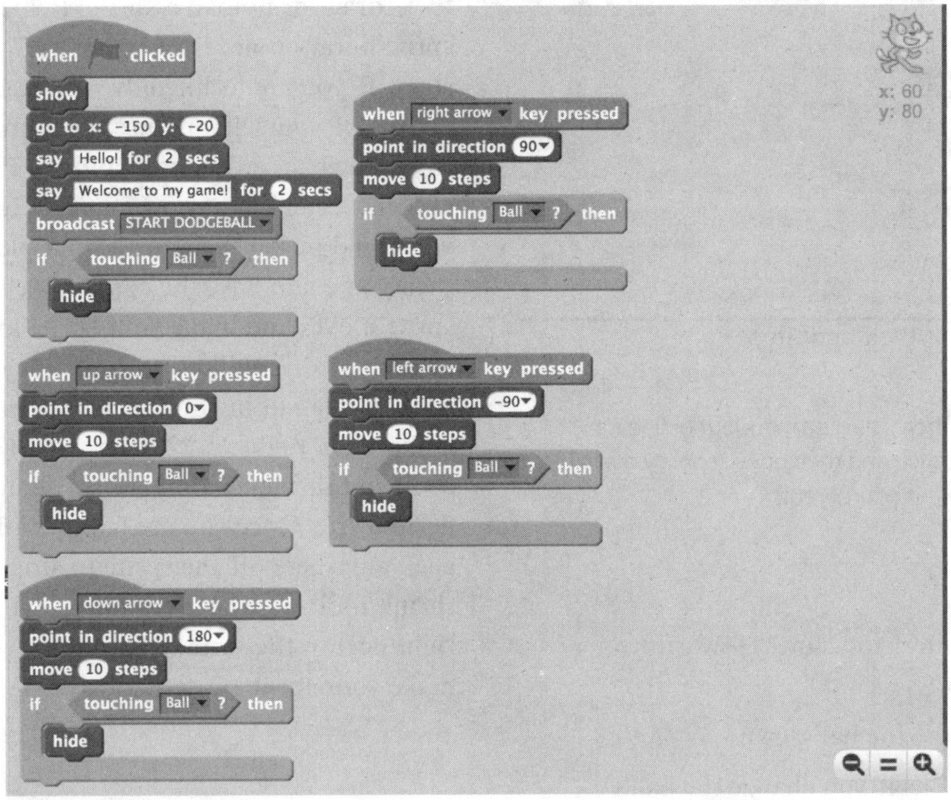

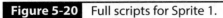

Figure 5-20   Full scripts for Sprite 1.

Step 12: The Ball Keeps Rolling ...

Your game is getting better! However, the ball is still rolling, and there is no one left to tag out! This is so because we have our ball moving in the "forever" loop. Don't worry, we can still stop something in the "forever" loop by putting the command "Stop this script" inside the "if/then" statement. This will tell our ball to move 10 steps until it touches Sprite 1. Then the ball will play a pop sound and stop. Refer to the script in Figure 5-21 to see the program. Your game is ready to play!

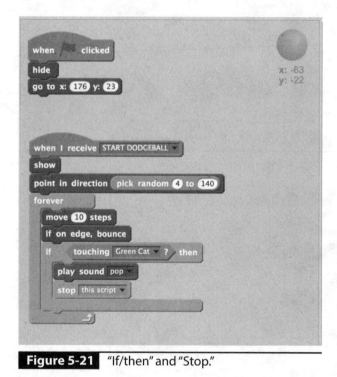

Figure 5-21   "If/then" and "Stop."

Classroom tips: There is a great handout online for parents about Scratch and their privacy policy, available at https://scratch.mit.edu/parents/.

Challenges

- Is your Sprite 1 too slow? How can you speed it up?

- Can you slow the ball down?

- What happens if you change the ball's random directions? Why do you think the random operator variables 4-140 in step 8 work better than say the variables 1-350?

- Using Broadcast Message, can you make a "Game Over" backdrop appear once Sprite 1 disappears?

Project 20: Scratch Maze Game

Cost: Free

Make time: 45–60 minutes

Supplies:

Materials	Description	Source
Computer	Computer with Internet access	—
(Optional)	(Optional) Raspberry Pi computer	SparkFun (DEV-13297)
Scratch	Scratch account	scratch.mit.edu

Step 1: Pick a Sprite and Shrink It!

Pick, draw, or upload a sprite! Remember, the sprite library is accessed under the "Scratch" stage. If you are feeling adventurous, take a picture of yourself, and use the costume editor to cut the background out and have yourself as a character in your own game! However, if you could pick a sprite with multiple costumes, then you could have the costumes change as the sprite moves and make your sprite appear to be animated. You'll want to shrink your sprite so that it can fit in your maze. The shrink tool is shown in Figure 5-22. Click on the shrink tool, and then click on your sprite until it is the desired size. Once you are happy with its size, click anywhere off the sprite to stop using the shrink tool. Alternatively, the growth tool is right next to the shrink tool should you want to make your sprite bigger.

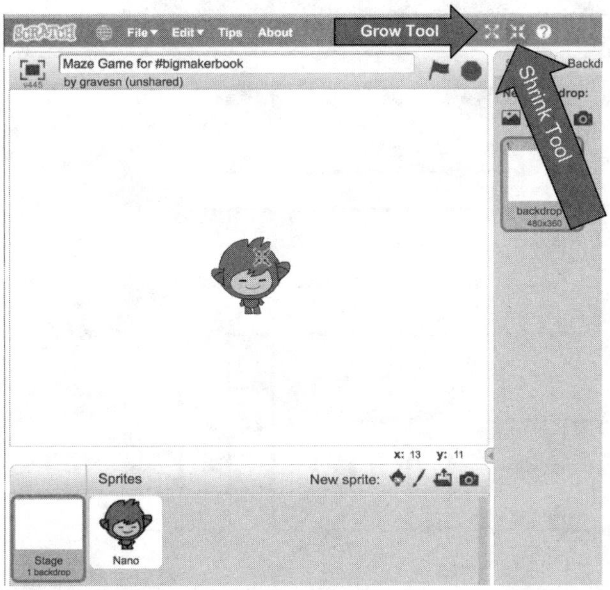

(Figure 5-24), you can erase a portion of the rectangle to be the end point of your maze. Use the line tool or rectangles, and design your own maze until you have something similar to Figure 5-25!

Figure 5-22 Shrink your sprite.

Step 2: Draw Your Maze

Click "Backdrop," and open the drawing tool. Use the rectangle to make a border for your maze, as in Figure 5-23. Using the toolbox

Figure 5-24 Toolbox.

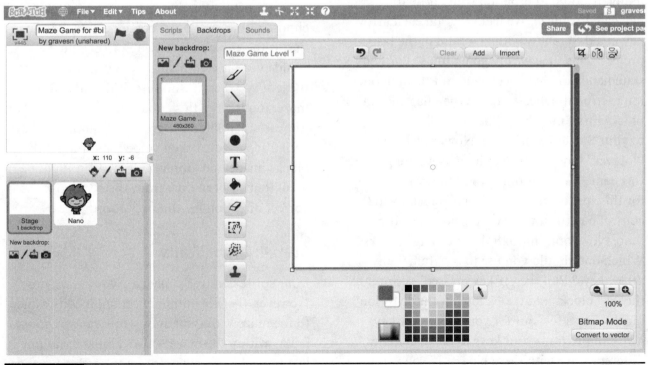

Figure 5-23 Outline maze.

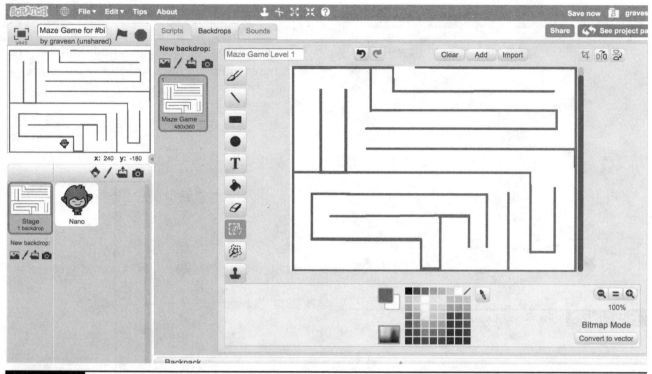

Figure 2-25   Ready for scripts.

Step 3: Program Sprite

Once you've got your maze designed to your liking, it's time to program your sprite to navigate the maze. We will program each arrow key in a different way than we did in the last project. This time we are going to use "if/then" statements for each arrow and put them in one long script attached to the "When flag clicked" block. First, pull the "When flag clicked" script to your Sprite 1 work area. Now we will need a "forever" loop because as long as we are playing this game, we want our arrows to work this way for this sprite. Drag an "if/then" block from the control palette, and place it inside the "forever" loop. Now look through the "Sensing" blocks. Which one should you use to program your arrow keys? Pull the "If key pressed" script into the "if" block, and take a look at the "Motion" blocks. Try a few out for your left arrow key. You can program your keys to move in many different ways with these blocks. If you want

your sprite to move left, you will need to change its x coordinates by a negative amount. I chose a sprite with multiple costumes so that I could also make it look animated as it moved. You can choose to put the "Next costume" block inside the "if/then" block or "Switch to costume." You could even have your sprite grow as it moves to the left, but then it wouldn't fit through your maze. Remember this code because you may want to use it later! Program all your arrow keys with a separate "if/then" block, but put all those blocks inside one "forever" loop as in Figure 5-26. Remember to program them to move the correct direction for their x,y coordinates.

Step 4: Maze Walls

Your sprite can freely navigate your maze now! However, there is nothing keeping it constrained to the maze walls you drew. How can you make it stay within the maze walls? Think about our last project and how we programmed the sprite

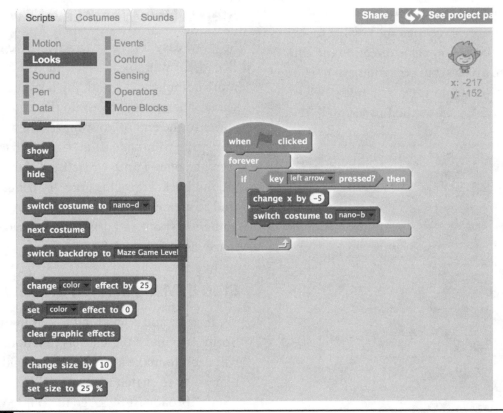

Figure 5-26 Left arrow and changing costume.

to disappear when we touched the ball. We can use a similar method to keep your sprite inside your maze. We need another conditional statement from the "Control" blocks. Drag an "if/then" block into your left-arrow script as in Figure 5-27. Using the "Sensing" blocks, we will tell our sprite that if it touches the maze walls, it will not be able to move. To do this, we just use the "Touching color" block, and put it inside the "if" statement. Then you'll need to program the sprite to go in the opposite direction from what it does when it isn't touching the maze wall. You might think that this would make the sprite run to the right, but since we programmed all the other keys and it won't be touching the wall anymore, it will actually just stop our sprite in its tracks. For fun, I am changing my sprite's costume to a frowning costume as well so that when it runs into the wall, it can't move and it frowns. Now you can duplicate this "if"

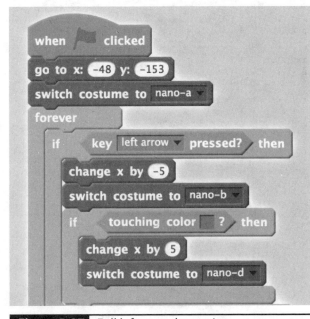

Figure 5-27 Full left arrow key script.

statement and nest it into your right-arrow script. Remember to change the x value to –5. You can also copy and paste direction for your up and down arrows, but you will have to change the "Motion" block to your "Change y by" block. Figure 5-28 shows the full script.

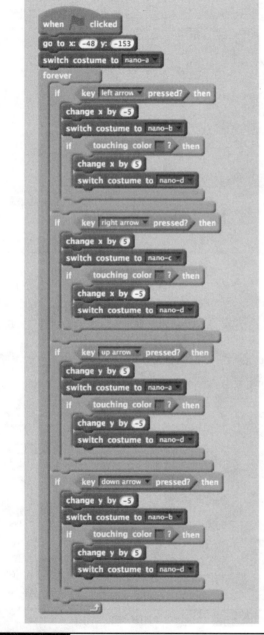

Figure 5-28   Full script for Sprite 1.

Step 5: Now What? Level Two?

See how easy it is to program a maze game? Because it was so easy, let's make it a little more complicated by adding a level 2 to your game. The easiest way to do this is to add a sprite to the end of your maze. We will program this sprite to broadcast a "Change backdrop" message once Sprite 1 reaches it. Let's create a second backdrop. This time, try using the circle tool to create a circle maze. You can draw circles and erase spots and draw lines to make your own maze, as in Figure 5-29.

Step 6: Make Scratch Switch Levels

We have to add something to our main script in Sprite 1 to make these levels function. Plus, you will need to make a new script to send Scratch a message to switch to level 2. First, add a "Start" broadcast message to your "When flag clicked" script underneath the "Costume" block, as in Figure 5-30. Now that we are sending a broadcast message, we have to also add a receive message. You need to tell the game that when it receives the "Start" message, it should switch to the level 1 backdrop. You'll also want to tell it how to know when it is time for level 2. If and when Sprite 1 reaches the arrow sprite, the backdrop should switch to level 2. You'll want to place this all inside a "forever" loop, as in Figure 5-31. You also need to tell Sprite 1 where to go when the backdrop switches to level 2. Grab an "Event" script and put the correct motion block for where you want your sprite to be when the level changes.

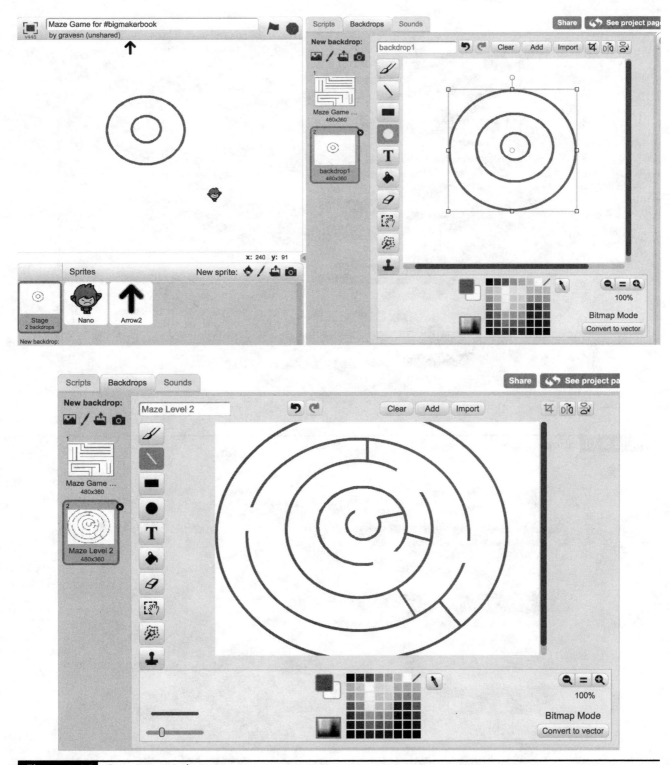

Figure 5-29 Drawing a circle maze.

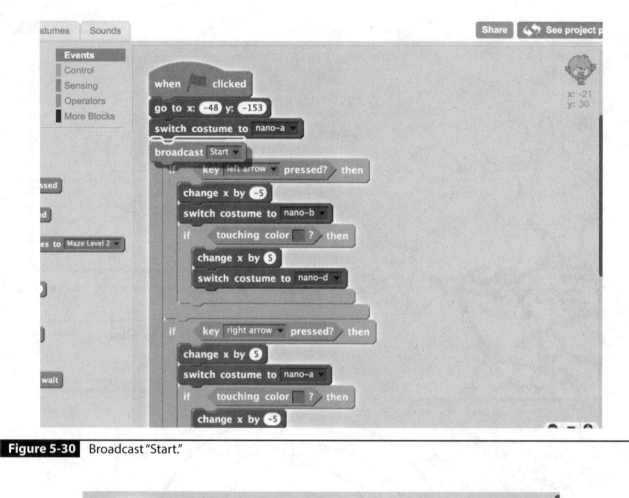

Figure 5-30 Broadcast "Start."

Figure 5-31 Start level 1/level 2.

Step 7: Hide Sprite 2

Unless you want to move the arrow to the center of your maze, you'll need it to hide when the level is switched. To do this, just click on the arrow sprite, and add the script "Show" to the "When flag clicked" script and a "hide" block to the "When backdrop switches" script, as in Figure 5-32.

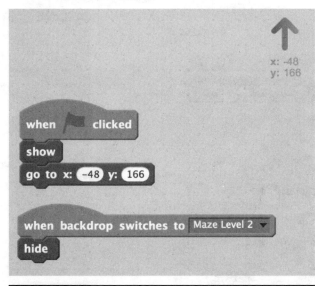

Figure 5-32 Hiding sprite.

Step 8: What Do I Win?

Let's give our sprite something to achieve once it reaches the center of the maze. I found a candy heart sprite and programmed it as in Figure 5-33 so that it will only appear on level 2 of the game. To make something happen when Sprite 1 reaches this candy heart, I'm going to switch back to the Sprite 1 scripts.

Step 9: Nom Nom Nom

Remember that "Motion" block "Change size by 10"? What if we use it to feed Sprite 1 when it reaches the candy heart? To do this, we will need to edit the "Event" block that only works on level 2 of our game. You should already have this script running, and it should tell Sprite 1 to go when level 2 starts. We are going to add an "if" statement inside a "forever" loop so that we can reward Sprite 1 when it reaches the candy heart! Inside your "if" statement you'll need a sensing "Touching sprite" block and some "Look" blocks to tell the player that the game is won! You type whatever you want your sprite to say

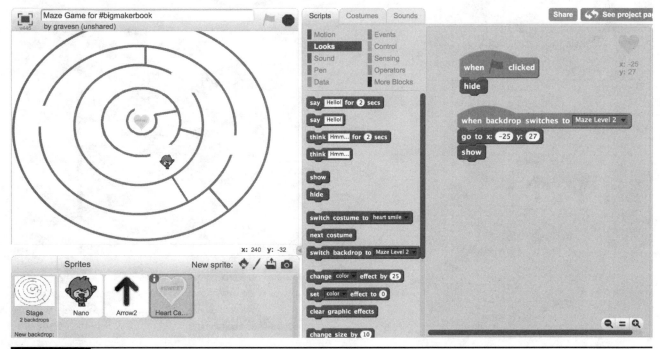

Figure 5-33 Candy heart sprite.

when it reaches its goal inside level 2. We added the "Change size by 30" block just for fun. What happens now when your sprite reaches the candy heart?

Step 10: Finishing Touches

We also thought it would be fun to add a little sound to tell the player that the game is over. You can record your own voice in the sound scripts by clicking the drop-down arrow inside the "Play sound" block, or you can click on the "Sound" tab in your workspace and record a sound there.

Add a new end game backdrop (Figure 5-34). We chose the "Rays" backdrop because it looks like it would make a great end screen! You can also type your winning message here by using the "Textblock" tool. Now, to program it to activate,

Figure 5-34 Adding an end game backdrop.

you need to make sure that you are working in Sprite 1's script area. You'll add a "Broadcast end game" from the "Event" blocks and place it inside your "if" statement underneath the "You won" sound because we don't want the screen to change until the statement is said (Figure 5-35).

We also need to create a "Receive message" or nothing will happen. You'll want to add a "When I receive end game" "Event" block and attach a "Hide" block from "Looks" and a "Switch backdrop to rays" block also from the "Looks" block (Figure 5-36). Now the game will switch to the end of the game!

Step 11: Tidy It Up!

You've made a great game! However, you almost forgot something! You will have to click on your "Arrow" block and make sure to add a "When backdrop switches" "Event" block and a "Hide" script just to make sure that it doesn't pop up somehow! You'll want to do this every time you add levels just to keep errors in game play to

a minimum. Also, you'll notice that once you add the "end game" broadcast, the sprite won't grow anymore. If you want your sprite to grow a couple of times, try adding a "Repeat" block like we did in Figure 5-37. In this way, our sprite

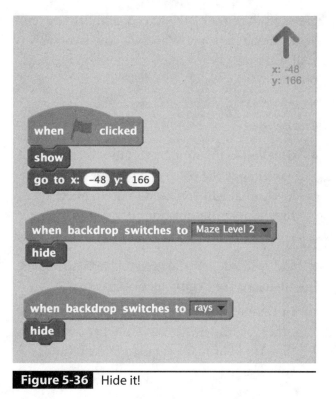

Figure 5-36 Hide it!

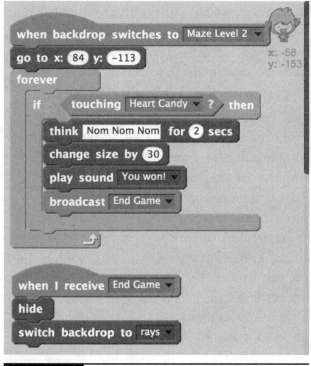

Figure 5-35 Broadcast end game.

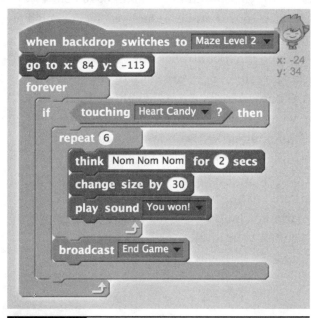

Figure 5-37 "Repeat" block.

grows to the edge of the screen, and then the end game screen pops up. What will you do now? Will you keep adding levels or move on the next project?

Challenges

- Can you keep adding levels to this game? If you change the color of the maze in the next level, what will you need to do to make sure that your sprite functions correctly?

- Can you add a chasing sprite?

- Can you add a restart sprite that sends Sprite 1 back to the beginning of the game?

- In Chapter 8 you will learn how to make a timer for Makey Makey. Could you add a timer to your maze game to make it more complicated?

Project 21: Arduino littleBits Project—Using Ardublock

You should experiment with Scratch until you start getting the hang of moving to the different palettes of coding blocks and learning the basics of that programming language before you move on to this project. Once you feel like you are ready to write *sketches* (what programs for Arduino microcontrollers are called), then this is a baby step to get you started. If you are an Arduino veteran, you still might be interested in using littleBits to test your sketches instead of hooking up wires and breadboarding lots of different components. littleBits is a pretty awesome tool because littleBits can be used for any manner of things. If you have this "Arduino" bit, you can even use littleBits to program your inventions.

Cost: \$–\$\$

Make time: 45–60 minutes

Supplies:

Materials	Description	Source
Computer	Computer with Internet access	—
Coding program	Arduino program and Ardublock	SparkFun
littleBits Arduino	littleBits Arduino (w6), Power littleBit (p1), Speaker littleBit, (o24)— Optional: other littleBits	littleBits.cc

Step 1: Get to Know littleBits Arduino

Arduino is an open-source platform with its own programming language and hardware. You can buy many different types of Arduino boards. In fact, we'll use two different Arduino boards in Chapter 7 and another in Chapter 9. This is what makes it a very handy programming language to learn because it gives you infinite possibilities when creating your own inventions with electronics. You can use your new knowledge of "if/then" statements to control the world around you, which is pretty amazing. You'll start looking at things in a whole new way because you can start to dissect how something is made and how you can make your own version of it.

We won't delve too deeply into Arduino itself because there are tons of other project books out there for you, and learning the whole programming language will take some time! For now, let's look at just the littleBits Arduino and using another graphical programming language called *Ardublock* to baby-step you into learning full Arduino coding/programming. The littleBits Arduino is unique among Arduino microcontrollers because you can snap together components rather than wiring them up. There are three direct inputs on this Arduino: bits d0, a0, and a1 (Figure 5-38). Don't forget that you

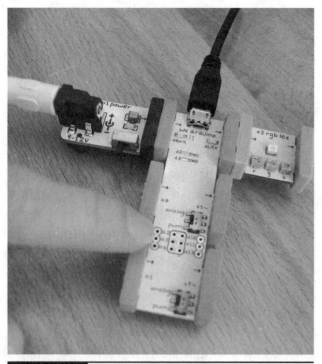

Figure 5-38 Other pins on the littleBits Arduino board.

will have to power your Arduino from this input side for your invention to function. For now, we are just going to work on creating blinking lights and are not going to delve into "if/then" statements and inputs. However, in the future, you could put different input bits on this side of your board and control your invention with an "if/then" statement just as you would control your sprite in Scratch!

There are even more pins located on the littleBits board, but if you are using those, then you probably aren't working through this beginner project! Pins are how you wire components from the outside world to your Arduino and control them with your computer program (known as a *sketch* in the Arduino world). These pins (which can be input or output) are connected to the actual legs of the microchip, and by connecting wires to these pins, you can create a path (or *route* for proper electronics terminology) for electrical signals to travel from the microchip to your wired

components—thus creating blinking LEDs, turning motors off and on, playing a tune, or even taking a reading.

Looking at the ArduinoBit, you'll see some tiny holes in the center of the board with numbers next to them. You can hook wires to these pins and assign functions. The other six pins we can use without soldering or wiring are located directly on the bitSnap connectors of the littleBit! This means that if you snap a LED bit into the top-right side, you'll be attaching that light to pin d1. So, if you write a program for d1 to blink off and on, then you'll see the light blink when you upload the program to the Arduino and power it up. The three pins on the right side of the ArduinoBit are outputs, and you program these pins by assigning scripts to either pin1 (d1), pin5 (d5), or pin9 (d9).

There is also a switch on the front of the ArduinoBit to convert your analog signal to a digital signal known as a *pulse-width-modulated* (PWM) signal. This will be helpful if you think your program isn't working. It might even be why! I ran through creating code for this project, and it didn't seem to be working at first. That's when I realized that I had my Arduino set to "analog." Plus, when you get into more advanced Arduino programming, this will be an integral part of the process. For this project, set your pins to "pwm."

This type of programming is very different from Scratch, so we are going to baby step our way into it by using Ardublock to understand the setup and syntax for writing code in this programming language. In this project, we will be writing a simple code to make three red-green-blue (RGB) leads (bit o3) flash off and on.

One of the coolest things about Arduino is that once you upload the code to the board, it retains the program until you upload another program onto it. This means that you can unplug it and use it to make robots or light up Sphero tunnels. Once you complete these Arduino littleBit blinking lights, take a look at Chapter 13 for some mashup makerspace fun!

Step 2: Download Arduino and Ardublock

Download the Arduino IDE from www.arduino .cc/en/Main/Software. You'll also need to download the Ardublock tool for this project. We downloaded our Ardublock add-on from the Hummingbird Robotics Kit site because it has blocks specific to it as well as littleBits blocks. Plus, it has great instructions for downloading this software, or you can download Arduino and Ardublocks from SparkFun at https://learn .sparkfun.com/tutorials/alternative-arduino -interfaces/ardublock, which also has great tutorials for downloading the software.

Step 3: Open Arduino Software and Ardublock

Once you have all the software downloaded, open it. It should open a sketch with today's

date. After you've written a few sketches (programs), it will open the most recent sketch you worked on. The first thing we need to do is familiarize ourselves a little with the work area. You'll have to choose the Arduino board you will be programming first. Go to Tools → Board → Arduino Leonardo. Plus, you'll also have to choose which USB port will be sending the program. Your correct port will only show up in the Tools menu if your USB is plugged in and your littleBits are powered on. So plug in the USB to your computer and your littleBits Arduino module. Make sure that you have a powerBit plugged into one of the left inputs. Since we aren't programming the input, it does not matter which input on the littleBits Arduino you snap the power into. Now go to Tools → Port, and you should see a USB serial port that says Arduino Leonardo. Once you select the board, there should be a dot next to it, and the port selected should show a checkmark next to it. Without these, you can't run your program because the software won't know where to send the code. Figure 5-39 verifies the setup. If you

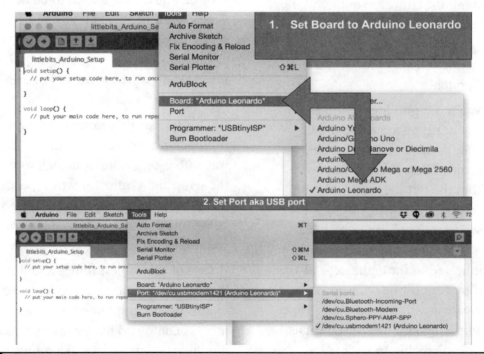

Figure 5-39   Setting littleBits Arduino board.

aren't sure which USB port is the correct one, unplug the board, and that port will be gone. You'll have to close and open the tools menu to refresh, but once you plug your USB port back in, you should easily be able to tell which one it is! Click on the correct port, and we are ready to start coding! Your littleBit Arduino is preprogrammed to blink, but we are going to look at what writing that program looks like and tinker with the code.

Step 4: Verify and Upload

Even though we are going to use Ardublock, it's still important to understand your work area. Every time you write some Arduino code, you'll want to check that code for errors. One of the great things about Ardublock is that it will help you to write your code correctly! In the Arduino environment (Figure 5-40), if you have a coding error, you'll get something like Figure 5-41. As a beginner, that error code won't totally help you to understand what you did wrong, but it will help you to get started debugging your code. For this project, you shouldn't get any errors

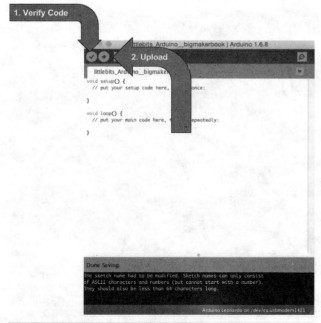

Figure 5-40 Arduino environment.

Figure 5-41 Error example.

here because Ardublock will help you to correct your coding errors before you upload the code to this sketch area. Ardublock is great at keeping error codes at a minimum. However, if you do not click your blocks together, you will get an error. Once your code is correct, you'll send it to the Arduino sketch area, and then you will click the upload arrow to send your program to your littleBits Arduino board. But we don't have any code yet, so let's get to it!

Step 5: Void Setup

Think of *void setup* like your sprite list in Scratch. You had to create different sprites in order for your game to function. In the same way, you have to program the "pins" on your Arduino board for your Arduino board to function. You do not have to program all the pins; this is why the software has you call out what pins you are using for your program. When you get further along with Arduino programming, you can also title your pin above the void setup so that you can more easily change your code in the "void" loop. You'll see some examples of this in Chapter 7. Your void setup is basically like building a road on the chip for the electricity to drive on. You only build it once.

Step 6: Void Loop

Once the road is built with your void setup, you can give the driving directions in the "void" loop.

The code you write here will allow the electricity to drive over and over and over again. This is why I say that the void "loop" is where the magic happens. Open a few example sketches and see if you can identify any patterns emerging. To open an example sketch, go to File → Examples. Look at a few sketches. What do you notice? (See Figure 5-42.)

You might notice the grayed out text that shows up after //. This is a handy way for programmers to write notes in their code because anything after the // does not affect the program. It is also a great way for you to get used to Arduino awesomeness by looking at expert programs, reading the notes, and running the programs! However, first, you'll have to understand a little more about programming pins. So let's program three pins with Ardublock and look at our code!

Step 7: Ardublock

To open Ardublock, go to Tools → Ardublock. You will get a window similar to Figure 5-43. It is handy to drag your Ardublock window over

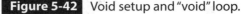

Figure 5-42 Void setup and "void" loop.

so that you can see what the blocks look like in Arduino coding. It will help to send you on your way to becoming an Arduino super genius! Inside Ardublock, you'll see a "loop do" block. This is where you will tell your Arduino what to do. The blocks you drag here will end up in the "void" loop in the Arduino sketch. Ardublock

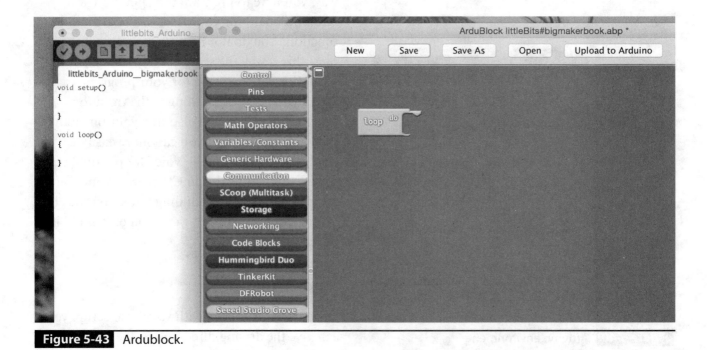

Figure 5-43 Ardublock.

will magically fill in your void setup for you when you place items inside the "loop do." This is one of the reasons it is a great starting point if you are teaching young makers about Arduino or even if you are new to Arduino yourself. It will build your road for you!

Step 8: Add LEDs to littleBits

Grab a couple of different bits, such as RGB LED, and Light Wire and a speaker. It will be fun to see what our Arduino code does with different bits, so keep them all handy, but attach a RGB LED to the d1 pin, one to d5, and one to d9 (Figure 5-44).

Step 9: Set Light to Blink ON/OFF

The first thing you have to do is set your light to come on by assigning the Arduino pin and telling it how you want it to function. To do this, drag a "Set digital pin" block from the "Pin" palette into your work area. It will automatically set the pin to 1 and the mode to "HIGH." This means that the Arduino will tell anything attached to pin 1 to flash on (Figure 5-45). Drag

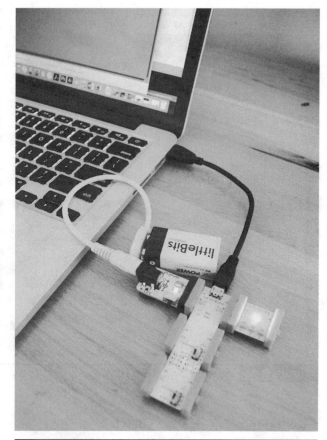

Figure 5-44 Arduino attached to computer and with LED.

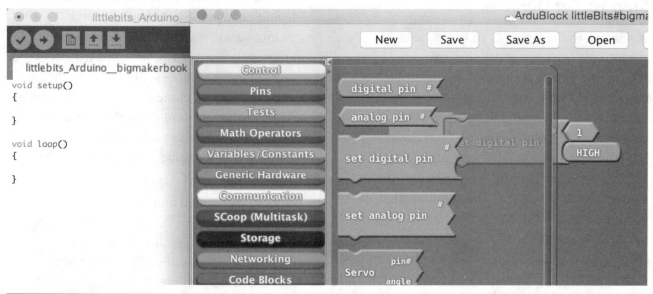

Figure 5-45 Setting digital pin HIGH.

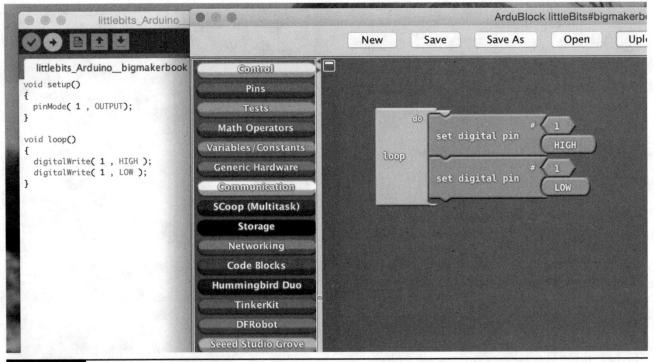

Figure 5-46 Setting digital pin LOW.

a second block of this and snap it to the first. Change the setting to "LOW" (Figure 5-46). This will tell the bit attached to pin 1 to have the lowest amount of power. So, if we attach a LED to pin 1, you would hope to see it light up and blink off. Try it, and see what happens.

Step 10: Set Duration

If you ran the program from step 8, your light might flash once, but it definitely wouldn't blink off and on. This is so because we have to tell our program what length of time we want the LED to be on and what length of time we want it to be off. In Arduino, this is called *duration*. We'll use this in Chapter 8 to create a song with correct note lengths! To set the duration of each function, drag a "Delay millisecond" block under the "Set digital pin HIGH" (Figure 5-47). Drag another to the LOW code, and click "Upload to sketch." You should see the Arduino code now in the Arduino IDE. From here on, you can upload it to the board and watch your

first set of lights blink. Our program is telling the component attached to pin 1 to blink HIGH and blink LOW. In other words, blink on and blink off.

Step 11: Clone

Since that was pretty easy, let's clone our code and change the length of the blink. In a moment, we will use this same cloning technique to program our other two pins. In Ardublock, you can clone the whole script by right-clicking on the top block or clone only a set of scripts by clicking where you want to start cloning. Let's clone these four blocks and change the duration of blink to half a second by changing 1000 to 500 as in Figure 5-48.

Upload this sketch and notice the coding in the Arduino language. Notice that every line of Arduino code ends with a semicolon. Also notice that when we tell our pin d1 what to do, we are using a "digitalWrite" command.

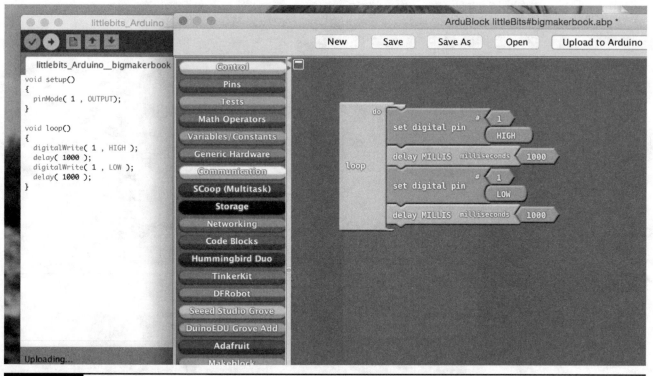

Figure 5-47 Setting the length of time for light to blink and upload sketch.

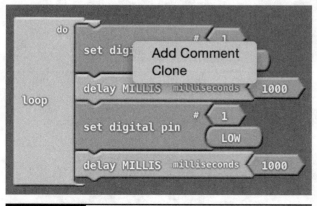

Figure 5-48 Cloning blocks of code.

This command only affects the information in the parentheses next to it and before the semicolon. The semicolon tells the Arduino that you are done with that line of instruction and it should move on to the next one. Without the semicolon, the computer won't know that that line of instruction is finished. When I first started tinkering with an Arduino, this is where I had most of my errors! I kept forgetting to put semicolons at the ends of my lines!

Also notice the duration of the blink is performed with the code "delay" and that the length of the "delay" is set inside the parentheses. The first blink ON/OFF is 1 second in length, and the second blink OFF/ON is ½ second.

Let's use the cloning technique to write the same script for the last pin on our Arduino littleBit. Clone the entire script, and change pin 1 to pin 9, as in Figures 5-49 and 5-50. Make sure that you have a light attached to the third bitSnap connector on the right side of the board. Upload the code to your littleBit Arduino, and watch your lights flash! (Figure 5-51). You'll notice that just like Scratch programming, the lights are not blinking at the same time; instead, the lights are flashing as the program runs from top to bottom.

One common problem I had when beginning Arduino programming was assigning the wrong pins and then wondering why my beautiful coding was not working. Look at Figure 5-52 and see if you can spot the difference in the two sets of code.

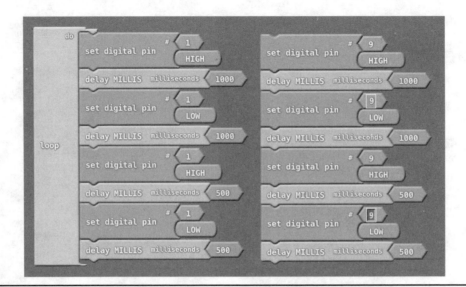

Figure 5-49 Cloning and adjusting a pin.

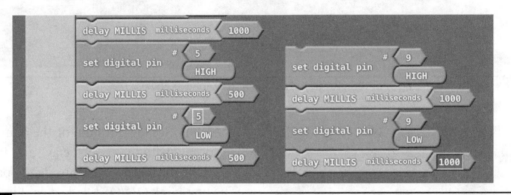

Figure 5-50 Clone for pin 9.

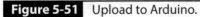

Figure 5-51 Upload to Arduino.

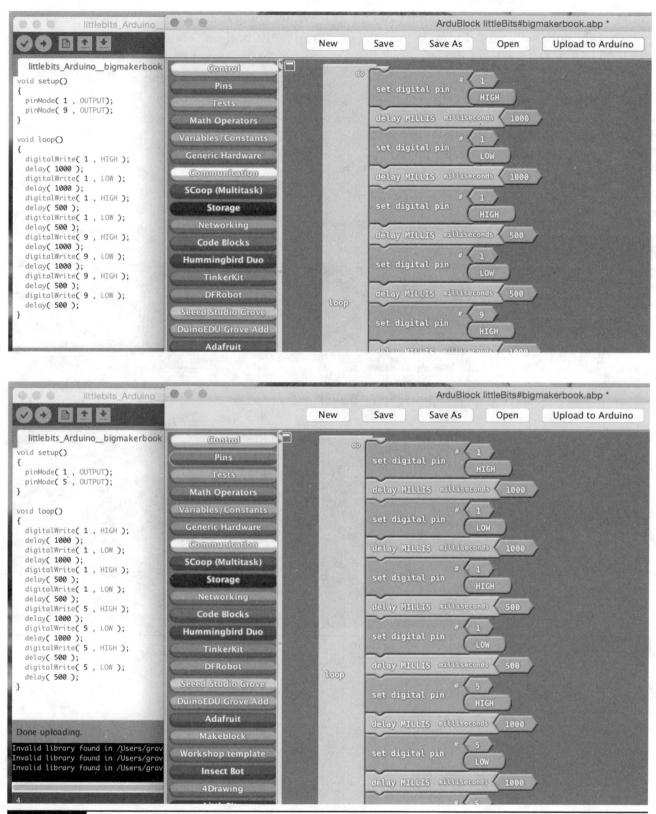

Figure 5-52 Can you spot difference?

If I were to run my program with my LED on pin 9 but the code says pin 5, I wouldn't see any rewarding blinking from my LED (Figure 5-53).

Step 12: Make It Blink!

Clone your entire script again, and make sure that it matches the second code in Figure 5-54.

We are going to have our lights blink from the top to the bottom (Figure 5-55). Try adding a few different output bits of littleBits to your Arduino and see what happens. What happens if you put a speaker on pin 1? What does the number bit do?

Figure 5-53 Full window of Ardublock and Arduino.

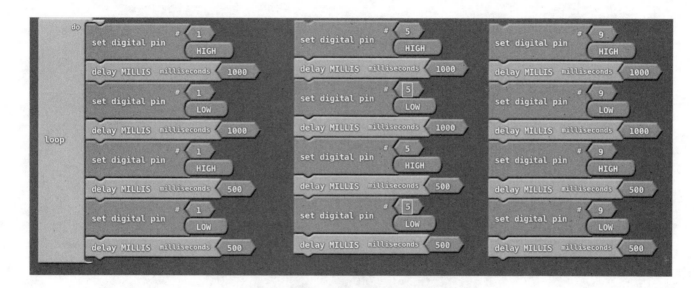

```
littlebits_Arduino__bigmakerbook

void setup()
{
  pinMode( 1 , OUTPUT);
  pinMode( 5 , OUTPUT);
  pinMode( 9 , OUTPUT);
}

void loop()
{
  digitalWrite( 1 , HIGH );
  delay( 1000 );
  digitalWrite( 1 , LOW );
  delay( 1000 );
  digitalWrite( 1 , HIGH );
  delay( 500 );
  digitalWrite( 1 , LOW );
  delay( 500 );
  digitalWrite( 5 , HIGH );
  delay( 1000 );
  digitalWrite( 5 , LOW );
  delay( 1000 );
  digitalWrite( 5 , HIGH );
  delay( 500 );
  digitalWrite( 5 , LOW );
  delay( 500 );
  digitalWrite( 9 , HIGH );
  delay( 1000 );
  digitalWrite( 9 , LOW );
  delay( 1000 );
  digitalWrite( 9 , HIGH );
  delay( 500 );
  digitalWrite( 9 , LOW );
  delay( 500 );
}
```

Figure 5-54 Full code Ardublock vs. Arduino.

Figure 5-55 littleBits blinking.

> ## "Coding" Challenge
>
> Now that you have a little coding under your belt, what will you create? Can you make a multilevel chasing game? Or multilevel ball game? Maybe you want to focus on Arduino and learn to program real objects? Make something awesome, and share your challenge project with the #bigmakerbook hashtag!

More Resources

- **Great coding resources for teachers:**
  https://code.org/educate/curriculum/teacher-led

- **Bitsbox:**
  https://bitsbox.com/

- **Wink Robot:**
  www.plumgeek.com/learn-to-code.html

- **Khan Academy:**
  www.khanacademy.org/hourofcode

- **Processing:**
  https://processing.org/

- **Ardublock:**
  https://learn.sparkfun.com/resources/tags/ardublock

- **Arduino:**
  www.arduino.cc/

- **Codebender:**
  https://codebender.cc/

Musical Instruments

THIS CHAPTER EMPLOYS everyday objects to create musical instruments that have science, technology, engineering, and mathematics (STEM) connections. The projects range from super easy to more complex as you progress through the chapter.

Project 22:	Popsicle Stick Kazoo
Project 23:	DIY Phonograph
Project 24:	PVC Pipe Organ
Project 25:	Rainsticks
Project 26:	One-String Guitar
Project 27:	Adding a Piezo Pickup and an Audio Jack to an Existing Project

Chapter 6 Challenge

"Musical instrument" challenge.

Project 22: Popsicle Stick Kazoo

These simple kazoos are quick to make, and they are a great introduction to the physics of sound. They create crazy, fun sounds, and you can feel the vibrations caused by the change in tone as you play.

Cost: Free–$

Make time: 10 minutes

Supplies:

Materials	Description	Source
Craft supplies	Jumbo craft sticks, assorted rubber bands, and straws	Craft or office supply store
Tools	Scissors	Craft store
Recyclable supplies (optional)	Backing from office labels, cereal bags, thin waxy paper wrappers, foil, and paper	Recycling bin

Step 1: Stringing the Bow

Take a large rubber band and wrap it from end to end on a craft stick, as shown in Figure 6-1. If you are working with younger children, pair them up so that one can hold the craft stick, and the other can stretch the rubber band.

Step 2: Insert Straws

Cut two 1-inch sections of straw. Place one straw under the rubber band 1 inch away from the end. Place the other straw on top of the rubber band about 1 inch away from the opposite end, as in Figure 6-1.

Figure 6-1   One under and one over.

Figure 6-2   Testing finished kazoo.

Step 3: Wrap the Sandwich and Test Kazoo

Place the kazoo in between your lips and blow. Does it make a sound? The kazoo will bow slightly due to the straws and rubber band. Try pressing gently with your mouth using different pressures to affect the pitch (or sound) of your kazoo. Now try blowing on the kazoo with different strengths for different effects. Changing the pressure should affect the pitch of the kazoo (Figure 6-2).

Challenges

- What other materials make a kazoo work best? Provide several options, including recyclable materials (wrappers, backing from office labels, cereal bags, etc.), and let students explore how the pitch is affected. Why do you think the pitch changes with each material type?

- Can you make the kazoo louder? Download a free decibel meter app, and see who can make the loudest kazoo.

Classroom tip: You can create another variable for students by providing different thicknesses of straws and by having them tinker with types of straws and moving the locations of the straws. You could even investigate what would happen with using multiple straws at one time. Ask students to predict how they think the sound will change as they place more and more straws between the popsicle sticks. Students can also experiment with different materials in the center of the kazoo. Remove the large rubber band that wraps around the center popsicle stick, and try using ½-inch strips of cardstock, label or sticker backing, waxed paper, or foil to replace the rubber band in the center. The strips won't need to wrap all the way around; they just need to rest in between the popsicle sticks with one end under one straw and the other end over the other straw. Place the top popsicle stick back on top, and hold the kazoo together with rubber bands.

Figure 6-3 DIY phonograph in action.

Project 23: DIY Phonograph

Make your own DIY record player out of paper, and learn a little about geometry and physics too (Figure 6-3). This super-simple project will allow you to make a working record player with just a few household supplies. No electricity required.

Cost: Free–$

Make time: 15 minutes

Supplies:

Materials	Description	Source
Craft supplies	Play-Doh or modeling clay, size 9 or 10 sewing needle, roll of paper	Craft store
Grocery supplies	2-inch piece of bamboo skewer, bendy straws, newspaper, and clear tape	Grocery store
Record	LP record 331/3 rpm record	Grandpa's record collection

Step 1: Make a Stand

Make a cone of modeling clay or Play-Doh on a table. Place a bamboo skewer upright in the middle of the clay cone, and press the clay firmly around the skewer, as shown in Figure 6-4. This will be where the arm of your phonograph will pivot.

Figure 6-4 Play-Doh stand.

Step 2: Sound Cone

For this build we used an 18- × 18-inch cone made of drawing paper. Part of the fun and experimentation with this project is changing the size of the cone and the type of paper. So don't be afraid to attempt this project with paper of a different size. To begin, make the paper into a cone by curling one corner in and twisting it into a point. Try to create a cone with as sharp a point as possible. Use clear tape to hold the cone together.

Step 3: Get to the Point

To stiffen the end of the cone and provide support for the needle, wrap a piece of tape around the tip of the cone a couple of times. Push the needle through the cone sideways about ¼ inch away from the end of the cone, as shown in Figure 6-5. Angle the sharp end of the needle away from the point of the cone. Push the needle point about ¼ inch out of the cone, as pictured in Figure 6-5. The needle will carry the vibration from the grooves cut into the record inside the cone, allowing you to hear the record play!

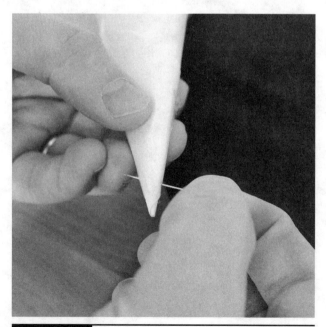

Figure 6-5   Needle placement.

Step 4: Attach Cone to Arm

The bendy-straw arm creates a pivot point that allows the needle and cone to travel in the grooves from the outside toward the inside of the record. If the bendy straw is positioned correctly on the cone, it will add just enough weight to keep the needle in the groove on the record. The bendy part of the straw will mostly likely end up around three-quarters of the way down the cone. Line the straw up with the needle, and use clear tape to attach it to the straw. Slide the straw onto the skewer, and test out the balance. Later you may need to adjust the position of the straw if your needle slides off the record too easily when you rotate it. It may also be necessary to add a straw to support the cone and keep it in the upright position.

Step 5: Rotation

Sharpen a standard No. 2 pencil, and push it snugly through the hole in a 33⅓ rpm record. Push the record up about ¾ inch. The record may stay in place, but if it doesn't, wrap a rubber band or tape around the bottom of the pencil to make it thicker. You may want to place a piece of paper under the pencil so as not to make any marks on the table when it rotates.

Step 6: Put the Needle on the Record

Put your pencil in the upright position, and place the needle gently on the record near the outside edge, as shown in Figure 6-6. As you begin to spin the record, you should begin to hear a hiss. Once the record begins to play, adjust the speed of your rotation so that the music and voices play at a normal speed. If the needle drifts quickly across the record or skips, you may need to shift the pencil closer or further away from the needle.

Figure 6-6　Needle and cone in position.

Challenges

- Can you make the record louder without using electricity?

- What could you design with a three-dimensional (3D) printer that would make this project more reliable and easy to use?

- How could you apply what you learned to other objects that produce sound?

- Does the type of paper used to construct the cone affect the sound (e.g., wax paper, drawing paper, poster board, etc.)?

Classroom tip: This project is a great way to apply math concepts such as volume of a cone, radius, and diameter. As a class, you could experiment with variously sized cones to see how they affect the sound. How does a long cone versus a short cone sound? How about fat versus narrow? Download a decibel meter app, and you will have a great source of data for comparison and drawing conclusions. Check out Chapter 11 for more applications for math with this project!

PVC Instrument Safety Tips

Polyvinyl chloride (PVC) is a readily available material for building percussion and wind instruments. PVC as a material demands respect, and you should follow all product guidelines and regulations for disposal for your area. Here are some general guidelines to consider when cutting, priming, and gluing PVC pipe.

1. Always wear goggles to keep PVC splinters and dust out of your eyes.

2. Gloves are essential for skin protection when using PVC primer and glue.

3. If you are sanding or cutting PVC, wear proper respiratory protection.

4. Always cut away from you, and follow instructions for using PVC cutters or saws.

5. Use primer and glue only in a well-ventilated area with proper respiratory protection.

Classroom tip: PVC projects do not have to be glued with PVC cement. You can drill a hole and use screws to secure the pipe or simply use a low-temperature hot glue. Gluing can always take place later, and if you want younger students to see the process, make a quick video. This will ensure their safety and allow them to experience how PVC is cemented together.

Project 24: PVC Playground Pipe Organ

Cost: $–$$$

Make time: 2–5 hours

Supplies:

Materials	Description	Source
PVC supplies	Two 2-inch-wide × 10-feet long PVC pipes	Hardware store
	Eight –2-inch PVC pipe traps (optional)	
	Eight 2-inch PVC unions	
	PVC cement and primer	
Construction supplies	Sandpaper, 220 grit, 8-inch zip ties	Hardware store
Tuner	Smart phone piano tuner app or tuner	Android Play store
		Apple App store
Paddles	A pair of flip-flops	Dollar Tree
Wood (optional)	2- × 4-inch scraps for cutting base	Scrap pile
Rubber bands (optional)	To keep cuts straight	Office supply store
Tools	PVC cutter or PVC saw, clamp (optional)	Hardware store

Figure 6-7 PVC pipe organ.

Step 1: Make a Plan

There are dozens of ways to set up a PVC pipe organ (Figure 6-7). Our example is going to include a few turns, but for the most part, it will be very basic in design. Many of the items in the list are optional because your design may require a different approach. We recommend that you browse the web for PVC instrument inspiration before building. Produce a list of design considerations such as the one that follows:

- How many notes do you want?

- Do you want to tune the instrument to a scale?

- What diameter of PVC will you use?

- At what height do you want to mount the playing surface? FYI: Some of the PVC pipes are over 52 inches in length in this build if you plan to use a straight piece of pipe.

- Will the instrument be freestanding or mounted to a wall or fence?

Step 2: That Hertz

Before you begin, you need to understand a little bit about how sound is actually made in this instrument. When you slap the end of the pipe with a flip-flop, the force of the movement causes the pipe to vibrate. The vibration pushes or pulls the air particles into pressure waves.

Depending on the length and width of the pipe, these waves will vibrate at different frequencies.

The *frequency*, or number of times a sound wave vibrates, determines the pitch that you hear. For example, the note middle C has a sound wave that vibrates 261.626 times a second. Hertz (Hz) is a unit that is used to determine the frequency of a sound wave; the frequency of a wave is the number of oscillations, or cycles, occurring during a 1-second period. Middle C has a frequency of 261.626 hertz. Charts of hertz values are readily available on the Internet. The values of the notes for this project are as follows:

Note	Hertz
C4	261.626
B3	246.942
A3	220
G3	195.998
F3	174.614
E3	164.814
D3	146.832
C3	130.813

Step 3: Tube Length

To calculate the length of your pipe using the following formula, you will need a couple of values to put input into the formula. First, input the diameter of your pipe in inches. We decided to use 2-inch pipe for our instrument. You will also need to input the speed of sound for your altitude in inches per second. We live close to sea level, so the speed of sound for our area is 13,397.244 inches per second. The speed of sound can fluctuate with temperature as well, so keep in mind that the formula in Figure 6-8 will provide a close approximation.

Don't let the math scare you. We will walk you through it. Let's start by handling the multiplication in the parentheses first. Look up the frequency in hertz of the note. We are going

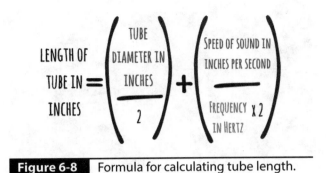

Figure 6-8   Formula for calculating tube length.

to use the note C3 as our lowest note. It has a frequency of 130.813 hertz, so plug that number into the formula. We are going to multiply the frequency in hertz 130.813 by 2.

Now it's time to divide the values left in the parentheses (Figure 6-9).

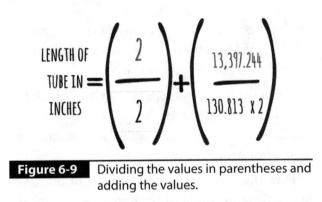

Figure 6-9   Dividing the values in parentheses and adding the values.

All that is left is to add the remaining values together to get the length of our pipe to create the note C3. Therefore, to create the note C3, we need a piece of 2-inch pipe that is 52.2 inches long.

Step 4: Check Yourself and the Speed of Sound

If you are creating a simple instrument with no turns, then all you need to do is cut the pipe to the length 52.2 inches to achieve the note C3. However, the speed of sound varies due to altitude, humidity, and temperature. You may want to cut your first pipe a little longer than this measurement to be on the safe side. To

Figure 6-10   Making the cut.

ensure that your cut is straight, you can wrap a rubber band around the pipe or use a miter box to guide the saw (Figure 6-10).

Use a piece of sandpaper or sanding block to remove any rough edges. At this point, we also marked the note and hertz value so that later we would not get confused.

Download a piano tuner app to your phone, grab a flip-flop, and test the tone of your pipe by slapping the end with a flip-flop (Figure 6-11). If it is flat, you will need to cut a little off the end to achieve the desired note. If you are sharp, you

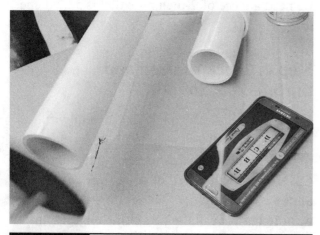

Figure 6-11   Checking the tone and the speed of sound.

will have to cut a new pipe that is longer. After cutting the first couple of notes, you will have an idea of how much to adjust your length to get the desired tone.

Step 5: Calculate the Curves

We were building our instrument for a playground for toddlers and small children. This created a bit of a challenge because most of them are not tall enough to reach the top of a 52-inch pipe. We drew out several plans and eventually decided on a plan with an even playing surface that was a good height for children ages 2 to 8 years.

Our design starts with a 10-inch section with a 180-degree turn at the bottom. This creates an even playing surface and allows the rest of the length of the pipe to travel back up the fence. To keep the weather out we added a 90-degree turn at the top of the pipe and a board to lay across the playing surface when it is not in use.

To calculate how much to cut off of each pipe to account for the 180-degree turn, we used a string to estimate the length of the curve. We estimated that it was about 8.5 inches around the inside of the curve and 11 inches around the outside. To test our estimate, we subtracted 8 inches from 53.2 inches and cut a piece of pipe that was 45.2 inches long. We downloaded a piano tuner app to our smart phone and fit the 180-degree curve to the end of the pipe. We tapped the end with a flip-flop and discovered that we were a little flat. We shortened the pipe a half-inch at a time. Our pipe finally sounded a good C4, and we knew that we needed to subtract 10 inches for the 180-degree curve.

We followed the same process to find that we needed to remove 4 inches for our 90-degree fitting at the top. We recommend that you cut conservatively and use our results as a general guide because PVC fittings can vary by manufacturer.

Step 6: Dry Fit

After you have cut and assembled your pieces, make sure that they are tuned. Mark the spot where each piece comes together and the note it belongs to. These markings will allow you to make sure that your pieces are pushed together at just the right spot so that your tuning does not change once it is glued (Figure 6-12). If you are using screws to secure your pipe, this is a great time to drill pilot holes through the fitting and into the pipe. PVC cracks easily, so before you insert screws, test your pilot hole size on a scrap piece.

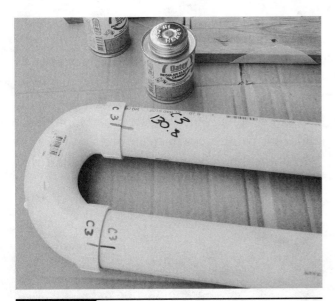

Figure 6-12 Dry fit.

Step 7: Seal the Deal

It's time to secure the pieces. PVC cement comes in two parts: a primer and cement. Most big-box supply stores have a kit that includes both. Make sure that all your cuts are free of debris and dirt, and work in a well-ventilated area. Wear safety goggles and gloves, and follow all safety precautions listed on the directions. First, you will need to prime the inside of the fitting and the outside of the PVC pipe. Make sure that the coating is even, and make a couple of passes if necessary.

Next, add an even coating of cement. Insert the pipe a little off the mark, and twist it into place. Use the marks you made earlier to ensure that all the pieces are connected correctly so that the tuning won't change. Allow the pieces to dry and cure in a well-ventilated area before continuing your work.

Classroom tip: If PVC cement is out of the question for you, there are other options. You can fit the pieces tightly and then drill a small hole through both pieces. Use a screw to secure both pieces and secure the pipe. PVC can split and crack easily, so be sure to test the screw size on a scrap piece. Another option is to hot glue the pieces and push them together tightly. You will have to work fast and keep an eye on your marks if you use this method.

Step 8: Get Attached

Our PVC instrument is going on a fence, and we found several different 2-inch pipe holders in the fencing and plumbing section of the hardware store that would work great. If you are on a budget, we recommend drilling holes 2 inches apart and using zip ties. Some quick research will reveal dozens of other mounting options, including some that use pipe fittings and 2-inch holes cut with hole saws. Think about what is available to you and within your skill set.

You can mount your organ directly to the existing fence or make it removable by using a 1 × 4 inch board. If you decide not to attach your organ directly to the fence, measure and cut a 1 × 4 inch board that spans the length between the fence poles. Be sure to add about 4 inches so that you can secure the 1 × 4 to the fence pole. Once you have cut the 1 × 4, mark the center of each end and draw a line down the middle of the length of the 1 × 4. If you are mounting directly to the fence, use a level to draw a line at the desired height you want to attach the brackets. Mark the center of the line you just drew and place the pipes in the mounts side by side on the board. Trace the location of the holes on the mounts; then drill pilot holes. If you are mounting directly to the fence it may be helpful to have a partner hold the pipe and mounting bracket while you trace the holes. Secure the pipes in their brackets to the 1 × 4-inch board or fence, using 1-inch or ¾-inch lag screws.

If you are planning on having your pipes off the ground, it is a good idea to add a brace to the bottom. Use some 1- × 4- or 1- × 6- and 2- × 4-inch scrap, and cut the pieces the same length as the mounting board. Use some 1½-inch wood screws to secure the scrap boards to the 2 × 4 in an L shape. Level the brace; then drill pilot holes on both sides of the 2 × 4 into the fence post. Use some 3-inch lag screws to attach the brace to the fence posts. To mount the board and pipes to the fence, drill two pilot holes on each side of the board, and attach it with 2-inch lag screws (Figure 6-13).

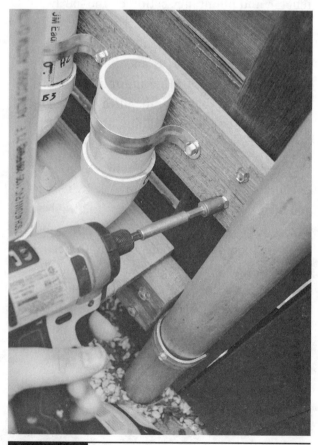

Figure 6-13 Attaching brace and mounting board.

Step 9: Paint

If you are going to paint your pipes, consider doing it after you work out how they will be mounted. You don't want to scratch an amazing new paint job while trying to figure out the best mounting method. Remove your pipes from the brackets, and use an aerosol spray paint that adheres to plastic. Lay them on cardboard, and paint one side at a time with thin, even coats. After two to three coats of paint on both sides, let the paint cure before attaching the pipes.

Step 10: Slapstick

You can go simple and take off your flip-flop and start playing a tune right away, or you can recycle an old Ping-Pong paddle or fly swatter. Whatever you decide to use, make sure that it will cover the pipe opening and that it has a soft, rubbery coating. We used a set of old flip-flops and some paint sticks to create our slapsticks. We cut the flip-flops into ovals and used hot glue to secure them to sticks (Figure 6-14).

Figure 6-14 Slapstick.

Challenges

- If possible, allow students to design a stand and PVC instrument for a real location.

- Can you create an artful and functional PVC instrument? Try using Google Sketchup to make a 3D model.

Shakers and Rainsticks

Project 25: Rainsticks

One of our favorite things about creating a rainstick is that the build can be very simple or sophisticated. Either way, you end up with a great product at really low cost. You can use readily available material such as cardboard with younger children, or you can use natural materials such as bamboo with older students. Take a look at the material list to get an idea of what you will need and which build will be best for you.

Cost: $

Make time: 1–2 hours

Supplies:

Materials	Description	Source
Recycling supplies	Thick cardboard tube, scrap paper	Recycling bin
Nails	Nails or screws ⅛ inch shorter than tube	Hardware store
Skewers	Bamboo BBQ skewers	Grocery store
Tools	Hammer, tape measure, ruler	Toolbox
Craft supplies	Yarn or scrap string, construction paper, craft glue, markers, duct tape	Craft store Scrap bin
Rain makers	Dry lentils or rice	Cupboard

Step 1: Spiral

We are going to create a spiral staircase for our materials to fall down. Many cardboard tubes already have a seam spiraling down them. If your cardboard tube doesn't have a spiral, you can easily create one by taping a string to the top of the tube and twisting it around. Many of the tubes that we used were made using a spiraling piece of 6-inch paper. So, if you use a string, just measure about 6 inches between each rotation, and use a piece of tape to hold it to the tube, as shown in Figure 6-15.

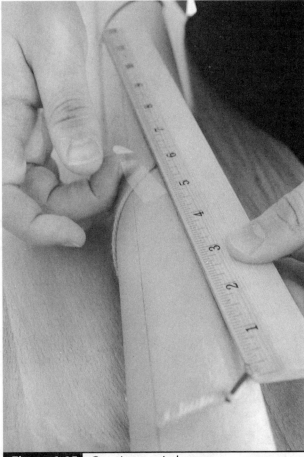

Figure 6-15 Creating a spiral.

Step 2: Mark the Spiral

In this step, we are going to hammer nails through the tube, but first we need to plan where those nails will go. Measure 2 inches down from the top of the tube, and place a mark every inch just above the spiral (Figure 6-16). Follow the spiral around until you are a few inches away from the bottom of the tube.

Step 3: Hammer Time

Tubes are not easy to keep still, so getting a partner to hold the tube steady makes this job go much faster. Before you begin, make sure that you and your partner have proper eye protection and that you have ample room to swing your hammer. Hammer a nail into each of the marks you made (Figure 6-17). Try to keep the nail straight, and avoid angling the nail so that it

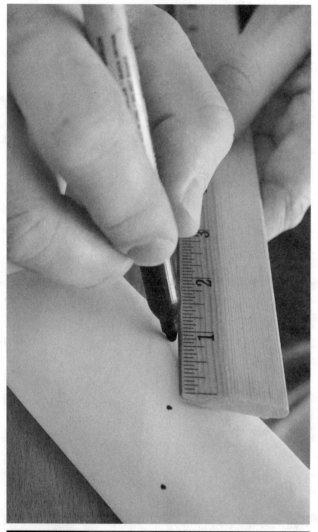

Figure 6-16 Marking the spiral.

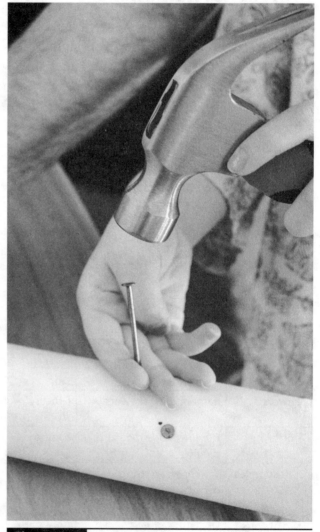

Figure 6-17 Hammer time.

pokes out the other side. Our tube was just over 2 inches wide, so we used 2-inch common nails so that the nail almost reached the other side. To determine the length of nails for your project, measure from the outer wall of your tube to the inner wall of the opposite side.

Step 4: Cap

If you are using a mailing tube, you might have been lucky enough to get a plastic cap for your tube. If not, a little duct tape and a rubber band will work just as well. It's tempting just to use duct tape to cover the end; however, remember that you will be adding material inside the tube, and you do not want it to adhere to the exposed sticky side of the tape on the inside.

Start by laying out about three strips of duct tape that are just over 6 inches long. Overlap each piece just slightly so you have about a 6-inch square. Next, repeat this step, only this time lay the pieces on top of your duct tape square sticky side to sticky side (Figure 6-18).

Trim any excess off the sides so that you have a duct tape square. Place your tube in the center of the tape square, and trace around it. If you want a uniformly edged cap, center a roll of tape

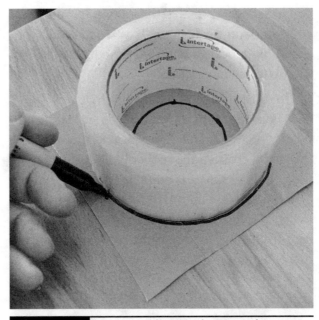

Figure 6-19 Center and trace a larger circle.

or old CD, and trace a large circle centered on the inner circle, as shown in Figure 6-19. If you want a cap that has some fringe, don't trace an outer circle; just use the square shape.

Use a marker to divide the square or circle by drawing lines from the center circle to the outer one. If you created the outside circle, then cut from the outside to the edge of the inner circle as shown in Figure 6-20. Repeat this step until

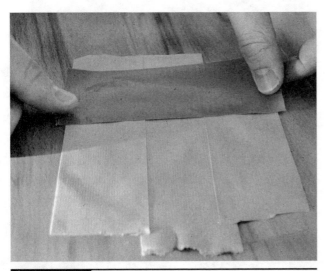

Figure 6-18 Tape square.

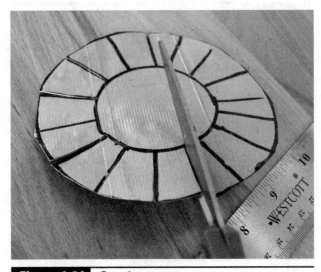

Figure 6-20 Creating rays.

you have several small strips radiating from the circle in a sun-ray fashion. If you decided to use a square shape, just cut from the edge of the square to the inner circle.

Hold the tube, and center the circle over the end. Fold the strips down over the side while placing a large rubber band around the edge of the strips. Overlap the rubber band a couple of times, and be sure it is on tight (Figure 6-21)! You can also use duct tape to cover the strips, but we don't recommend that you do this to both ends until you are sure of the type and quantity of the material you want to use inside.

Figure 6-21 Securing end.

Step 5: Make It Rain

After you have sealed one end of the tube, it is time to start experimenting with the type of material you want to use inside of your rainstick. Each material will react differently, creating a unique noise. We recommend that you start off with a cup of dry rice, dry beans, and some marbles. Start by adding only one material at a time and sealing the end before you tilt the rainstick. Listen to the type of sound created by each material, and keep note of how

long your rainfall seemed to last. Try different combinations and quantities of filler until you reach the perfect rainfall sound for you. This is also an excellent time to add more nails if your material is falling too fast. Try going back and adding a nail at every half-inch.

Step 6: Seal and Decorate

Once you are finished experimenting, it is time to decorate the rainstick and ensure that none of the nails fall out. Some simple and quick ways to accomplish this are using construction paper and glue or wrapping the tube with duct tape. Measure and cut the construction paper so that it overlaps slightly. Coat one side with school glue, and wrap it around the tube. Continue this process until the length of the tube is covered and the nail heads are no longer exposed. Get creative and add stripes or patterns with cut paper and glue or markers (Figure 6-22).

Challenges

- You may have discovered that your rainstick only produces a few seconds of rainfall depending on the material inside. How could you modify the rainstick to make the sound last longer?

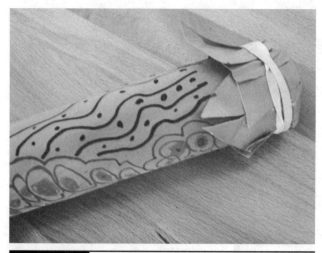

Figure 6-22 Decorated rainstick.

- Can you make a 3-foot rainstick that makes sound longer than a 4-foot one?

- How can you change the volume of your rainstick?

- Can you predict how the diameter of the tube will affect the duration of the rain? How would more or fewer nails affect the rain?

- Do you think the size of the material will affect the duration of the rain?

Classroom tip: A way to save time is to use a ruler to mark every inch before you wrap the string around the tube. The string can be used again by other groups after the tubes are marked.

Project 26: One-String Guitar

The *diddley bow* (one-string guitar) is a great starter instrument for anyone who is interested in building or playing a stringed instrument. Originally, a diddley bow was just a piece of wire that was tied tightly between two nails on a board. To put the string under tension, a glass bottle was added under the wire. The bottle elevated the strings like a guitar bridge and also amplified the sound of the instrument. For this build, we are going to use a cigar box or sturdy cardboard box instead of using a glass bottle to increase the volume (Figure 6-23).

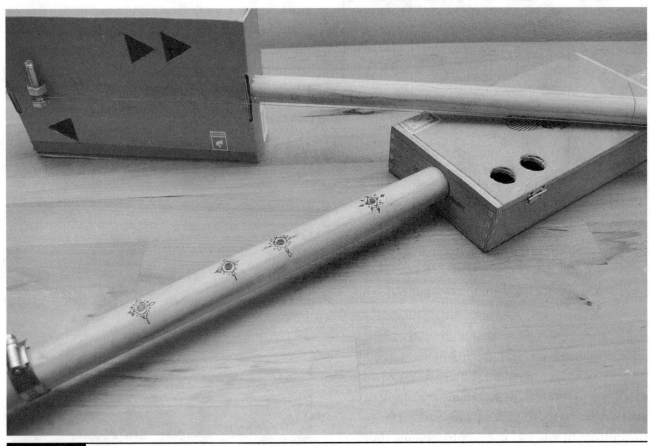

Figure 6-23 Diddley bow.

Cost: $–$$

Make time: 30–45 minutes

Supplies:

Materials	Description	Source
Guitar neck	Broom handle or 1- to 1¼-inch-wide wooden dowel	Recycle broken handle Hardware store
Guitar body	Cigar box or cardboard box 6 × 9 × 2 inches	Recycling
Tools	Electric or hand drill, craft knife, box cutter, file	Hardware store
Drill bits	⅛-inch drill bit, paddle drill bit, or hole saw that matches the diameter of wooden dowel, 1- to 2-inch paddle bit, or hole saw for cutting sound holes	Hardware store
Hardware supplies	No. 8 screw, ¾ inch, large bolt, steel pipe, hinge or handle for bridge, 1½-inch-diameter hose clamp, brass washers size No. 4S, variety of sandpaper	Hardware store
Guitar string	D guitar string	Musician's Friend Recycle broken strings

Step 1: Cut and Core

Measure and mark 32 inches, and place the wooden dowel in a clamp to cut it. Be sure to wear safety goggles. Clean up any splinters with sandpaper or a file.

Locate a suitable box that is about 6 inches wide, 9 inches long, and about 2 inches tall and preferably has a hinged lid. Figure 6-23 provides examples of suitable boxes. In most cases, if you are using cardboard or a very basic cigar box, the bottom of the box is going to become the top of the guitar. You can change this if you really like the design on the cigar box, but if you are using a cardboard box, you will need to use the

bottom as the strongest part of the box. Plus, if you decide to add a piezo pickup, you will easily be able to access the inside of the box if you build your diddly bow this way. We are going to drill or cut two holes on each side of the box so that the neck will rest just below the floor of the box. This will allow vibration to transfer from the neck to the box, where it will be amplified.

Next, measure the short end of the cigar or cardboard box, and mark the center on the top and bottom of the box. Connect the two marks so that you have a center mark.

You can usually determine the thickness of the bottom by measuring the thickness of the top or sides of the box. Most cardboard box and basic cigar box lids are an ⅛ inch thick, so for our box, we will measure down ⅛ inch. Next, we need to determine the radius of our neck. Measure the diameter or distance across the dowel, and then divide by 2 to determine the radius. For example, if you have a dowel that is 1 inch in diameter, the radius will be a half-inch. Measure this distance from the mark you made to account for the lid thickness. Repeat this step on both sides of the box. This should give you the exact center for your holes (Figure 6-24).

Figure 6-24 Measuring and centering holes.

Figure 6-25 Paddle bit.

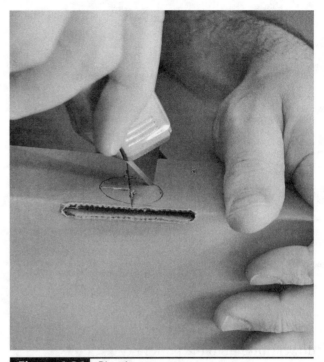

Figure 6-26 Pie pieces.

If you are using a cigar box, you will need a paddle bit the exact diameter of the dowel. When using a paddle bit, always clamp the box on a table, and be sure you use proper eye safety. Drill the hole with the paddle, and repeat the process on the other side (Figure 6-25). Clean up any splinters or rough edges with a file or sandpaper.

If you are using a cardboard box, center the dowel on the mark you just made, and trace a circle around it. Make a cut from the center of the circle to the edge of the circle to create pie slices. Then cut the outside diameter (Figure 6-26).

Step 2: Test Fit

Try to keep the hole just large enough that the dowel fits snuggly. Make a mark on the neck at 2 inches from the end, and slide the neck through the holes so that it lines up with the mark (Figure 6-27). If your cardboard box has a lip on the side of the lid, you will need to trim

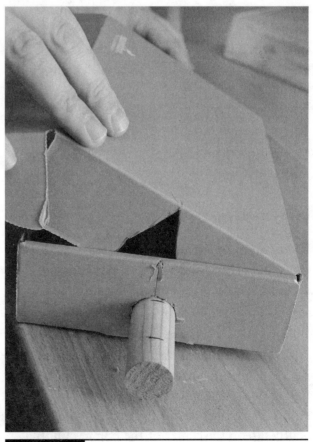

Figure 6-27 Test fit.

a wedge out so that it will close. If your hole is too large, you can cut some cardboard braces and glue them in place with hot glue. Whatever you do, be sure that you can remove the wooden dowel rod for now.

Step 3: Strings Holes and Holders

It is time to make a hole to hold the ball end of the string in place at the bottom of the guitar. The side that you made the 2-inch mark on will become the tail end of your diddley bow neck. Remove the dowel rod, and make a mark on the tail end 1 inch away from the end of the dowel rod. Make sure that you are wearing proper eye safety, and clamp or hold the dowel rod firmly; then drill a ⅛-inch hole from the top of the dowel through to the bottom.

Now it is time to turn our attention to the top side of the neck and the screw that will hold the string tight. Measure and make a mark 2½ inches from the top of the neck. Make sure that this mark is in line with the hole you made for the string. Use the ⅛-inch bit to drill a pilot hole through the rod at the 2½-inch mark. Sand the dowel smoothly after all holes are drilled.

Place a ¾-inch No. 10 screw in the hole, and turn it until there is about ¼ inch left before it meets the wood.

Step 4: Attach the Neck and Rivet

If you made the holes too large, place the dowel through the box, and use a hot glue gun to secure it by placing a bead of glue around the dowel. You may need to cut added cardboard for support.

A pop rivet placed in the hole at the tail of the neck will keep the neck from wearing down and failing due to the vibration and tension of the string. Place the pop rivet in the top hole; then gently tap on the center of the pop rivet until it comes out (Figure 6-28).

Figure 6-28 Pop rivet.

Step 5: String It

Slide the hose clamp over the end of the neck, and let it rest near the cigar box. We will get back to this part later, but it is essential that you have it in place before we tighten the string.

The diddley bow is shorter than a normal guitar, and it is a great way to give old strings new life. You can use an old guitar string or a new D string from a guitar. Flip the guitar over, and slide a small washer over the end of the string; then thread it through the bottom of the hole and through the pop rivet resting in the top of the hole. The washer will provide a larger surface for the string to rest, and the rivet in the top will prevent it from digging forward into the wood (Figure 6-29).

Hold onto the end of the string, and pull it as tight as you can get it by hand. Loop it around the screw twice. Turn the screw down until it is tight, and the string is secured. The screw should hold the string in place, and you should be able to strum it and get a sound.

Figure 6-29 Brass washer.

Step 6: Nut and Bridge

Make a mark 5 inches from the top of the neck.

Now it is time to transform the hose clamp into a guitar nut. Move the clamp up the neck, and rest the string on top of the large, smooth part of the clamp. Slide the clamp toward the 5-inch mark from the top of the neck, and then tighten the hose clamp (Figure 6-30). This smooth surface will elevate the string, increase the tension, and allow the string to vibrate more.

Your bridge can be made of anything over ¼ inch high and between 1 and 5 inches long. The bridge helps to elevate the strings and also transfers the vibration to the cigar box below. This is a good time to recycle things such as pieces of short metal pipe, old hinges, or a large bolt and nut. If you are using a bolt and nut

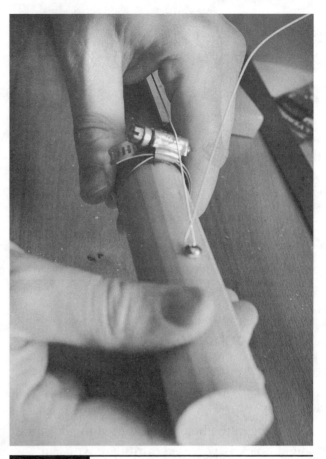

Figure 6-30 Slide clamp into position.

on a cardboard box, remember to tighten the nut so that most of the pressure rests on the neck and not the box. For cigar boxes, this does not matter because they are made of wood or compressed cardboard and are stronger. Slide the bridge under the string and toward the back of the cigar box. This is an opportunity for you to tune the open note for your diddley bow. Use a guitar tuner or download a guitar tuner app to your smart phone. Move the bridge piece up and down until you have tuned your string to a suitable note (Figure 6-31).

Figure 6-31 Bridge.

Step 7: Sound Holes

You have no doubt strummed your diddley bow and are ready to share its unique sound with the world. To let those sound waves that are resonating in the box out, you need to add some sound holes. If you used cardboard, you can make circular holes using the method described in step 1. You are not limited to circles with cardboard, so get creative!

For cigar boxes, you can make one big hole using a hole saw, or you can make four smaller holes using the paddle bit. When planning your hole placement, avoid getting too close to the edge of the box or the neck (Figure 6-32). If you are using a drill, we recommend that you remove the string while drilling.

Step 8: Mark a Scale and Frets

Measure the distance between the center of the bridge and the nut. This is the *scale length* of the diddley bow. Divide the distance in half, and make a mark. This should be roughly the twelfth fret position (Figure 6-33). A glass or metal guitar slide placed close to this mark will produce a note an octave higher than when strumming the open string. You can also gently rest a finger

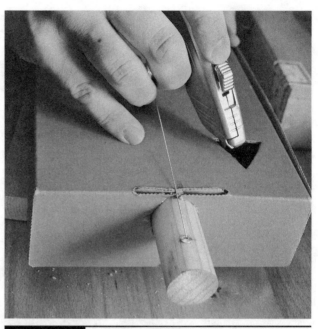

Figure 6-32 Sound holes.

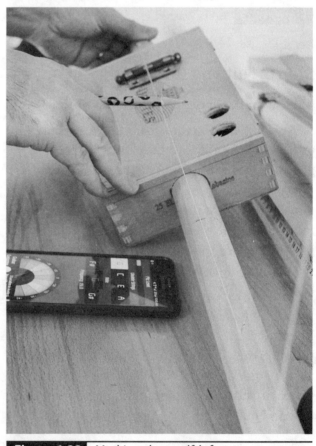

Figure 6-33 Marking the twelfth fret.

slightly above the string and pluck it softly. This action should produce a high-pitched bell tone called a *harmonic*. A tuner also comes in handy!

To mark the fifth and twenty-fourth frets, you will need to divide the scale into fourths. You should be able to find a harmonic at roughly one-quarter and three-quarters of the scale. Divide the scale of your diddley bow by 4. The result of this division will give you an approximate location of the fifth fret at one-quarter of the scale. Multiply one-quarter of the scale by 3, and you will get the location of the twenty-fourth fret at three-quarters of the scale. Make a mark, and test your measurement with a harmonic or tuner (Figure 6-34).

Figure 6-35 Finding other notes.

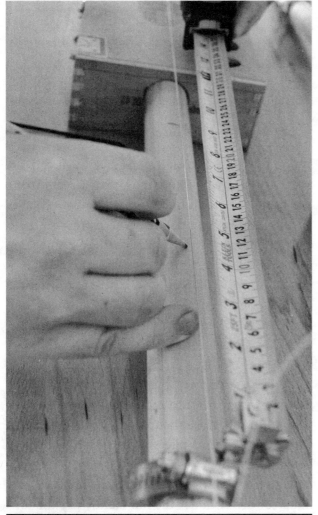

Figure 6-34 Marking the fifth and twenty-fourth frets.

The seventh and nineteenth frets can be found by dividing the scale into thirds at one-third and two-thirds, respectively. Divide the length of the scale by 3, and the result is the location of the seventh fret at one-third of the scale. Multiply this result by 2 to get the position of the nineteenth fret at two-thirds of the scale. Place a light mark, and test your location. You can also use the tuner and a guitar slide to find other notes and mark them on the neck (Figure 6-35).

Classroom tip: This is a great opportunity to teach fractions and engage students in a discussion about harmonics. You can use a string and divide it in halves, quarters, thirds, and eighths.

Step 9: Decoration and Fret Markers

The simplest way to denote your fret markings is with a pencil or permanent marker, but after spending so much time making an instrument, you can really make your diddley bow shine with some cool customization. If you decide to remove the string, mark the position of the nut and bridge. The nut and bridge must be placed in the same location for the fret positions to remain the same. Try your hand at a custom paint job, or use decorative furniture tacks hammered into the side of the neck to denote the fret locations.

Challenges

- How does the physical volume of the cigar box affect its loudness?

- Do more holes in the box make it louder? Find the most effective size and placement for sound holes by using a decibel meter to compare builds.

- What things other than boxes could be used to create a diddley bow?

Project 27: Adding a Piezo Pickup and an Audio Jack to an Existing Project

Cost: $

Make time: 15 minutes

Supplies:

Materials	Description	Source
Piezo pickup	Piezo transducer 6.35 millimeter/¼ inch (Radio Shack 273-073 or SparkFun SEN 10293)	Radio Shack SparkFun
Audio jack	Mono open-circuit audio jack	Radio Shack SparkFun
Solder	Lead-free solder	Radio Shack SparkFun
Soldering tools	Soldering iron station, fume extractor, third-hand tool	SparkFun Mouser Electronics
Tools	Drill, ⅛-inch drill bit, pliers	Hardware store

Step 1: Free the Piezo Element

If you purchased a piezo transducer, you will need to start by breaking the piezo element free. We are interested in the ceramic element inside because it possesses the ability to create an electric charge when it is flexed. Piezo elements are supersensitive, so the stress caused by vibrations from the string will cause it to release a corresponding electric signal. When that signal is connected to an audio jack and then to an amplifier, we can hear it. If you were able to get your hands on just a piezo element, skip ahead to step 2. It is very important to understand that piezo elements are somewhat fragile, and it is important to avoid bending or compressing them. Use a pair of pliers to break the tabs off the sides (Figures 6-36 and 6-37). Try to tear enough plastic away so that you can get a small flat screwdriver under the lid. Gently pry the top of the transducer off.

Figure 6-36 Using pliers to break tabs.

Figure 6-37 Broken tabs.

Once the top is off, turn it over and locate the small hole on the back of the transducer. Try to gently push the element out with the flat side of a bamboo skewer or matchstick. Exercise caution, and be sure not to bend or push too hard on the piezo element.

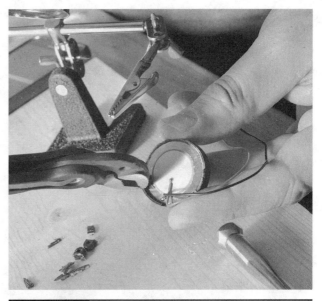

Figure 6-38 Breaking the sides away.

If it does not budge, try carefully breaking the sides away, taking care not to bend or crimp the piezo element. After breaking the sides away, try pushing with a skewer again (Figure 6-38).

Step 2: Wired for Sound

Your piezo element will have a red wire and a black wire coming from it. It's a good idea to think about how far away your element will be from your audio jack. If you need more room, simply solder on a few inches of wire to achieve the desired distance to the jack. In most cases, the 4 inches that are already soldered on the element will be ample (Figure 6-39).

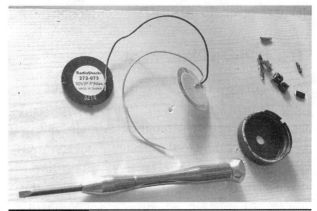

Figure 6-39 Piezo element.

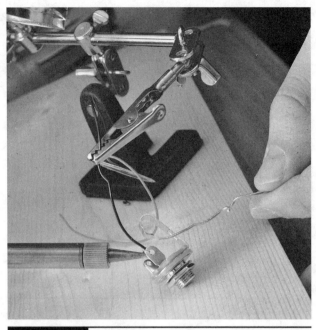

Figure 6-40 Soldering to the open jack.

A simple open-circuit jack has only two connections to which to attach wires. The black wire will attach to the tab that will make contact with the body of the audio cable once it is installed. Solder the red wire to the tab that will make contact with the tip of the audio cable (Figure 6-40).

Step 3: Hole

Drill or cut a ³⁄₈-inch hole in the side of the guitar or diddley bow. Remove the washer and nut, and place the audio jack through the hole. Add the washer and nut, and firmly tighten the jack into place (Figure 6-41).

Step 4: Piezo Placement

You will discover that where you place your piezo element will affect the tone of your instrument when it is amplified. Attach some modeling clay tape to the element, and try out some different locations. Try having a portion of the element fixed to the instrument while the other half hangs out into open space. You will

Figure 6-41 Placing the jack.

probably notice how sensitive the piezo is to the slightest vibration. To buffer the element from slight vibrations, we are going to place a small layer of foam on the back. When you are ready to commit on a location, use the old top of the piezo transducer to cut a piece of foam about the size of the element.

Place a small drop of hot glue on the foam, and stick it to the backside of the element. Finally, glue the piezo to its new home, and rock on! See Figures 6-42 and 6-43.

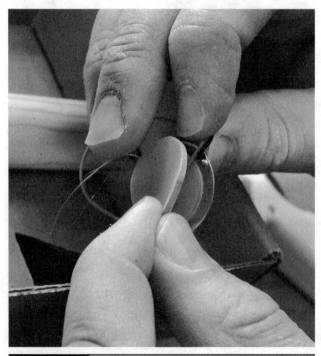

Figure 6-42 Gluing the piezo.

Figure 6-43 Piezo in place.

Challenges

- How can you get the warmest tone and sustain it from your instrument?

- What happens if you change the thickness and density of the foam?

- What other applications can you think of for using a piezo?

"DIY Musical Instrument" Challenge

Think about how we've used different types of materials throughout this chapter to make vibrations that create musical notes. Can you discover other materials that would make a great DIY instrument? Can you change the length or tension of your materials to make different notes?

After you've invented your own DIY instrument, tweet it to us @gravescolleen or @gravesdotaaron or tag us on Instagram and include our hashtag #bigmakerbook to share your awesome creations. We will host a gallery of your projects on our webpage to share with the maker community.

Sewing Circuits: Beginner to Advanced E-Textiles

You have experimented with paper circuits, done some beginner programming, and now it's time to sew these worlds together with conductive thread! These projects range from simple to complex. The more advanced projects assume that you have some working Arduino knowledge and are interested in testing that knowledge with threads of steel!

Project 28:	Sewing LED Bracelets
Project 29:	Wearable Art Cuff with DIY Switch
Project 30:	Heavy-Metal Stuffie with LilyPad Arduino

Chapter 7 Challenge

E-textile challenge: design your own stuffie.

Project 28: Sewing LED Bracelets

Even if you are a master at sewing, you need to start small when sewing electronics. I had LilyPad and Flora projects sit in my workshop for years before I finally sewed together a Firefly Jar Kit from SparkFun (Kit 11833). That was the moment I really fell in love with sewing circuits, and I've reused those pieces over and over again to teach myself how to get more complicated in my circuitry! Sew together this simple wrist cuff bracelet as your first wearable

project, wear it, and then reuse all your materials to make something even more complex and sparkly!

Cost: $$

Make time: 2–3 hours
(for beginner sewing students)

Supplies:

Materials	Description	Source
E-textile (workshop supplies)	For a classroom makerspace: E-textile Basics Lab Pack (SparkFun Lab 13165)	SparkFun
E-textile supplies (individual supply)	Five-pack of LilyPad LEDs SparkFun (white: DEV 10081; blue: DEV 10045; pink: DEV 10962; yellow: DEV 10047; red: DEV 10044)	SparkFun
Conductive thread	Conductive thread bobbin (30 feet) stainless steel SparkFun (DEV 10867) or Adafruit Stainless Thin Conductive Thread (Product ID 640)	SparkFun Adafruit
Battery holder	3-V LilyPad battery holder (switched) SparkFun (DEV 11285) or SparkFun (DEV 08822)	SparkFun
Battery	Coin cell battery CR2032	Amazon SparkFun

(continued on next page)

Materials	Description	Source
Bracelet body supplies	9- × 3-inch felt pieces, a regular thread assortment, metal snaps, metal clasps, or Velcro	Craft or fabric store
Sewing notions and supplies	Sewing needles, pincushions, scissors	Craft or fabric store
Embellishment notions	Assorted semitransparent or frosted buttons, beads, sequins, felt pieces, a small jewelry kit, as shown in Figure 7-1	Craft or fabric store
Ribbon (optional)	Transparent ribbon to hide LEDs and sewing (if desired)	Craft or fabric store
Insulators	Stretchy fabric glue or hot glue gun	Craft or fabric store

If doing this workshop with kids, precut many different colors of felt to this size.

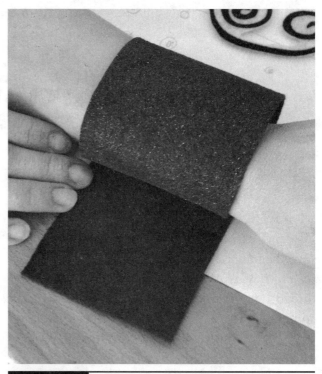

Figure 7-1   Check bracelet width.

Step 1: Check Bracelet Width

Prepare your bracelet first. Choose a color, and make sure that it is long enough to wrap around your wrist and overlap about 1 to 2 inches. We are going to protect the battery pack in that overlap, so it is essential that you don't cut your bracelet too short (Figure 7-1).

Step 2: Design and Diagram

Design your bracelet on paper first. Decide what you want the overall bracelet to look like aesthetically, and then decide where you want your lights to shine. I decided I wanted my bracelet to mimic a portion of Van Gogh's *Starry Night*, so my LEDs would be the stars in the night, and I found some swirly felt I wanted to trim and layer on top of my blue felt background. Figure 7-2 is the rough sketch I used to brainstorm this project. I also decided I wanted the design to cover the entire bracelet. Once your design is ready to go, it's time to determine your circuit routing.

Classroom tip: If you are doing this as a workshop with multiple students, the design element is essential. Some students may sit around saying, "I don't know what to make." Get them sketching or even just laying out elements on felt to get their creative juices flowing. I've had workshops where no one knew what to make and others where all the students had design ideas when they stepped in the door! Whatever you do, don't skip this step. This helps to give your students a drawing board and a reference to look back to while sewing.

Step 3: Lay Out a Simple Circuit or Parallel Circuit and Test with Alligator Clips

Because this is your first wearables project, you may want to just keep your circuit simple, but that would mean only adding one LED. Plus, adding a few LEDs and sewing a parallel circuit aren't too difficult. Figure 7-3 shows you how to lay out a parallel circuit and connect to the battery pack. Figure 7-4 shows you how to create a simple circuit bracelet. Lay out your LEDs on your sketch, and use alligator clips to test your

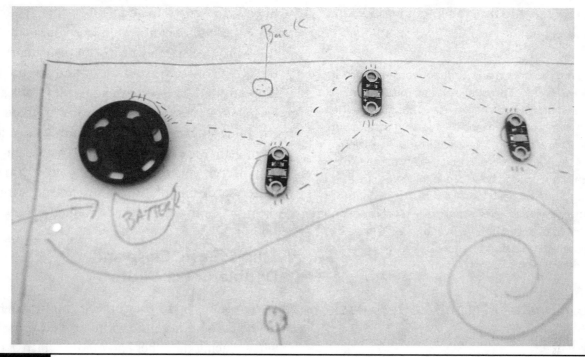

Figure 7-2   Design on paper.

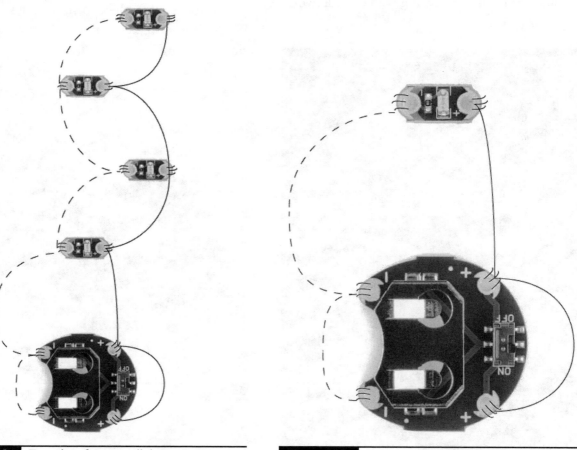

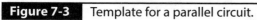
Figure 7-3   Template for a parallel circuit.

Figure 7-4   Template for a simple circuit.

circuit. Make sure that you lay all positives in the same direction!

Classroom tip: Since many students are new to circuit diagrams, I find that physically placing the LEDs on their sketch and drawing the negative and positive lines helps them not only to understand their sewing pattern but also how to read an electronic schematic. By placing the electronic elements down, you help students to physically understand how to read that two-dimensional drawing. Plus, you need to make sure that students have placed all their LEDs facing in the same direction. If students want to add notions over their LEDs, have them test those with alligator clips now, before they get much further along in this project (Figure 7-5).

Step 4: Baste LEDs and Battery Pack

Baste your LEDs and battery pack the old-fashioned way with a few stitches of regular thread, or use a dab of fabric glue to keep them in place until you start sewing with conductive thread. To *baste* is to add a quick stitch to keep an object in place until you place your actual stitch. You will simply tie a knot of regular thread that matches your fabric and sew a few loops around the positive and negative holes to keep your battery pack from falling off. You can clip these stitches off when you are done or leave them if they are unnoticeable. If you use glue, you just put glue on the back of your LEDs and battery pack and place them where you want them to be on your finished product!

Step 5: Baste Clasp and Double-Check Width

Baste your clasp to your bracelet on the backside of the fabric. Check the width around your wrist when placing the other end of your clasp. Also, make sure that your battery will be hidden under the extra fabric, and make sure that you do not sew your clasp too close to your battery pack. We don't want to accidentally short out your circuit (Figure 7-6).

Figure 7-5 Test LEDs.

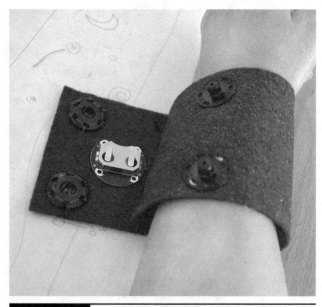

Figure 7-6 Basting clasp and battery pack placement.

Step 6: Sew Negative Traces

It's time to break out the conductive thread! Excitement! Remember to look at your circuit route from your drawing and the template in Figure 7-3. Sewing to the wrong place or skipping a LED will mean troubleshooting later. Cut off about an arm's length of conductive thread, thread your needle, and make a loop at one end. (Do not tie both ends together because conductive thread is too sticky to sew with this method. Instead, knot only one end, and leave a tail about 12 inches long. As you sew, you'll have to shorten that tail to keep it from getting stuck in the bracelet.) Let's sew from left to right and start by sewing the negative trace (routing) on your LEDs all the way to the battery pack on the right of your bracelet. Bring your needle up through the negative pin on the first LED. Make tight loops by coming up through the center of the negative pin and coming down right at the edge of the LED. Or you can sew from the outside in, but an important tip is that you keep your thread very tight and very close to the flower on the LED. Loop the thread through this pin three to four times, always making sure that you hold the thread tight so that you will get a good connection. (See Figure 7-7 for examples of a good stitch and Figure 7-8 for examples of a bad stitch.)

Classroom tip: Stay close during the first stages of sewing. I've had some students who thought it would be best to tie on the LEDs, and they made knots at the LEDs with their conductive thread, and then their circuit wouldn't light up. Also make sure students see examples of good and bad stitches before sewing.

Once you feel like your LED is secure, use a basic running stitch to get to the negative pin on the next LED. Loop three to four times through the center of the negative pin and to the outside as you did before. Continue your running stitch to each LED, and make sure that you loop

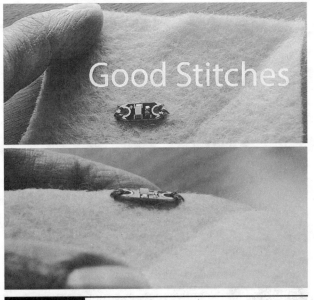

Figure 7-7   Good stitch examples.

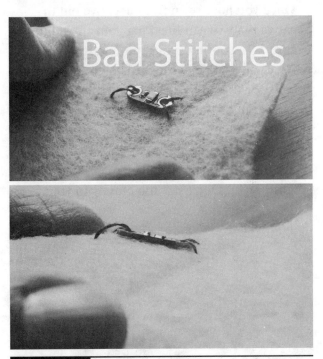

Figure 7-8   Bad stitch examples.

tightly around each LED. If you look closely at your LilyPad LEDs, you will see that the conductive element at the positive pin looks like a small flower. Make sure that you are making a good connection with this flower as you loop your conductive thread around the negative pin (Figure 7-9).

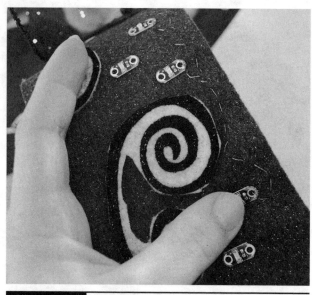

Figure 7-9 | Negative route as a running stitch.

Once you have sewn from negative pin to negative pin without breaking your thread (as in Figure 7-9), continue your running stitch to the battery pack on the wrong side of your bracelet. Loop your thread five to six times on the left negative pin of the battery pack; then sew a running stitch in an arc away from the battery pack to the right negative pin, loop five to six times, tie a knot, snip, and you are grounded (Figure 7-10). Don't sew too closely to your

battery pack; you do not want to accidentally short your circuit by touching the conductive thread to the battery pack in the wrong spot.

Classroom tip: Conductive thread is a bit tricky to work with because it isn't as slick as regular thread. It seems to stick and knot up easily. Keep your thread length shorter so that you don't get tangled up and have to start over. As much as I prefer all things SparkFun, the thread from Adafruit is actually a bit easier to work with than the SparkFun variety. You may want to consider getting a spool of thread from Adafruit, but the SparkFun LilyPad LEDs are my favorite.

Step 7: Sew Positive

Because we are already at the battery pack, let's start our route to positivity there. Cut a length of thread from one shoulder to arm's length. Thread your needle, and tie a knot at the bottom end. Come up through the positive pin in your battery pack at the far end of the bracelet, make five to six loops to secure your circuit and keep your battery pack from falling off, and then arc a quick running stitch to the other positive pin on the battery pack (Figure 7-11). Hand stitch

Figure 7-10 | Sewing negatives to the battery pack.

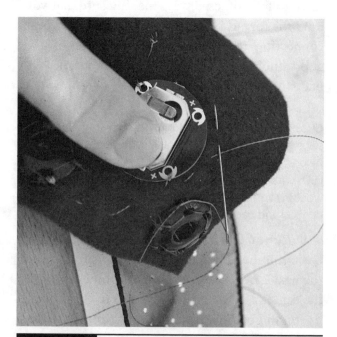

Figure 7-11 | Sew positives to battery pack.

a running stitch to the positive pin of the fifth LED, loop three to four times around this LED as you did before, and work backwards until you have all LEDs connected on one route (your positive trace). As long as you don't cross your negative or positive thread traces, you can get creative with your stitching like in Figure 7-12.

At your last LED, bring your thread to the wrong side of your fabric. Bring your needle up through the last stitch, and tie a knot as in Figure 7-13. Cut your thread close to the knot.

Figure 7-12 Getting creative with stitching.

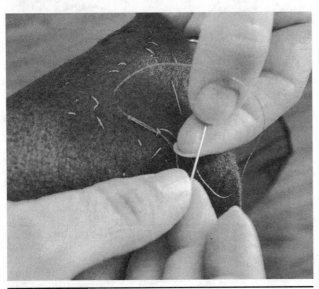

Figure 7-13 Tie a knot.

Step 8: Test

The moment of truth! Did you cross lines while stitching? Did you sew a straight route to all your LEDs? Let's find out. Place your battery in the battery pack and see your bracelet light up!

Did it not light up? Check your battery placement. Did you put it in backwards? Did one LED not light up? Check to make sure that you sewed both traces to each LED. I completed a bracelet once and skipped a whole LED on the positive trace (path)! This is an easy fix. You should be able to add the circuit routing to your LED as long as your opposing parallel line does not get in your way. Simply loop to the previous LED, sew a running stitch to your missed LED, loop through the LED three to four times, and sew a running stitch to the next LED. Did that work? Hooray? No? You'll want to check to make sure that you don't have any positive stitches near your negative stitches. Recheck your stitching to make sure that you used short, neat stitches. Long stitches have a tendency to reach places they shouldn't and cause a short circuit. Also, check to make sure that you sewed very close and tight loops around your LED pins. See Figures 7-7 and 7-8 for examples of good and bad stitches.

Step 9: Embellish

I wanted to hide my stitches and components, so I layered a ribbon on top of my bracelet. Plus, I thought it made it look even more like the nighttime sky. Embellish your bracelet with ribbons, notions, or stitches, and you are almost done!

Step 10: Sew Clasps

When you have your bracelet looking the way you want, remove your basting stitches, and sew on your final clasps. It's time to wear this DIY bracelet with pride (Figure 7-14). Remember to

Figure 7-14 Finished bracelet.

sew a clasp; you'll have to sew the male clasp on the backside of your project and the female clasp to the correct side. Also remember to measure along your own wrist before sewing.

Challenges

- You don't have to sew your parallel circuits in parallel lines. Can you make a working spiral parallel circuit?

- What other shapes can you create with parallel circuits? Can you make a star?

- What other shapes lend themselves to parallel circuits?

- How many LEDs can you add before losing current?

- Can you add dimensionality to your project by adding shapes with felt? If you do this, where will you put your LEDs?

Project 29: Wearable Art Cuff with DIY Switch

Once you've learned all your circuits with the paper circuit projects in Chapter 5, you can start to take that knowledge into your wearable projects. We are getting a little more complicated in our circuitry with this project. Since you learned to make your own DIY switch in Chapter 2, we are going to continue to hack projects with our own DIY switches. This "Starry Night" cuff uses a DIY switch with notions you can find at your local craft store. Who needs a battery pack with a switch when you can make your own? Plus, in this project, you'll learn to embellish your wrist cuffs with hand-stitched embroidery techniques to recreate your favorite artwork. Instead of covering the entire cuff with art, we decided to keep the focus on top of the wrist cuff, so you can have your own miniversion of Van Gogh's *Starry Night* to stare at throughout the day and take daydreaming to a whole new level (Figure 7-15).

Cost: $$

Make time: 2–3 hours

Figure 7-15 "Starry Night" stitch design template.

Supplies:

Materials	Description	Source
LilyPad LEDs	Five-pack of LilyPad LEDs (yellow: DEV 10047)	SparkFun
Battery holder	LilyPad coin cell battery holder (no switch): DEV 10730	SparkFun
Battery	Coin cell battery CR2032	SparkFun
Conductive thread	Conductive thread bobbin (30 feet) stainless steel SparkFun (DEV 10867) or Adafruit (Product ID 640)	SparkFun Adafruit
Microcontroller with LEDs (optional)	LilyPad Protosnap/LilyTwinkle (includes a battery pack, LEDs, and preprogrammed microcontroller to make LEDs twinkle like fireflies; it is perfect for this project) SparkFun (DEV 11590)	SparkFun
Microcontroller à la carte (optional)	LilyTwinkle (or you can buy the twinkling microcontroller à la carte and decide to use it in your next project, if it won't work in this one) SparkFun (DEV 11364)	SparkFun
Carbon transfer sheet	For transferring stitch design to your fabric.	Sublimestitching.com Back of this book
Embroidery template	Starry Night Stitch Design (Figure 7-15).	Back of this book
Bracelet body	Metal snaps or metal clasps, 9- × 3-inch plush felt, faux fur, or fleece pieces, regular thread assortment	Craft or fabric store
Sewing notions and supplies	Sewing needles, pincushions, scissors	Craft or fabric store
Embellishments	Assortment of embroidery floss (we used gold, yellow, and shades of blue)	Craft or fabric store
Insulators	Clear fingernail polish, stretchy fabric glue, or hot glue gun (you might want to glue LEDs before sewing as tacking/basting stitch but also because these materials are nonconductive, you can use them to fix short-circuits)	Craft or fabric store

Step 1: Measure Length

Like the last project, you want to measure your bracelet length first. Remember, we want the cuff to be long enough to wrap around your wrist and also to overlap about 1 to 2 inches. The battery pack will be housed in the overlap, so it is essential that you don't cut your bracelet too short! (Go back to Figure 7-1.)

Step 2: Design and Diagram

You will want to lay your cuff out folded as if you are wearing it to decide LED placement, as you did in Project 28. (You can also refer to Figure 7-16.) Go ahead and transfer the stitch design with a carbon transfer sheet for the embroidery stitching step so that you can make

sure that you place your LEDs in a "twinkly" spot. If you feel confident in your embroidery skills, you can skip the transfer and eyeball your stitches or draw them on with sewing chalk.

Figure 7-16   Folded cuff.

Classroom tip: Share Van Gogh and Monet paintings with your class. Let them look through art books for design ideas and inspiration. The Impressionist painting strokes can be mimicked with embroidery stitches. Feel free to use our "Starry Night" template or design your own wearable art inspired by these impressionist artists.

Step 3: Lay Circuit Out and Test with Alligator Clips

As we did in the preceding project, you'll also want to lay your elements out on a sheet of paper and draw your parallel circuit. Although, this time, instead of having your negative stitches, go to the battery pack; you will stitch your negative route to the female metal clasp. The male metal clasp will complete the circuit because you will connect this clasp to the negative pins on the battery holder. See this circuit laid out in Figure 7-17 and my hand-drawn circuit routing in Figure 7-18. You'll connect the positive route as you did before, beginning with the battery pack and ending at the last LED.

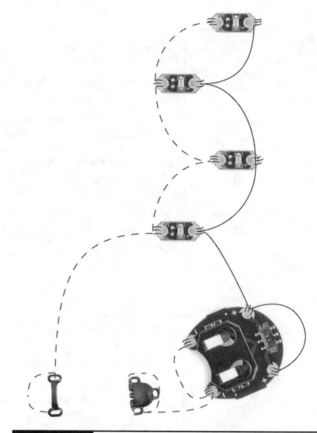

Figure 7-17   DIY switch circuits handout for students.

Figure 7-18   Hand-drawn circuits example.

Now let's test your metal clasps and make sure that they will work as a DIY switch. Connect one alligator clip to a negative pin on the battery pack and the other end to the male clasp. Connect another alligator clip to the negative on one LED and to the female clasp. Use a third alligator clip to connect the positive pin of the LED to the battery pack. Now see if your clasps will function as a switch to turn on your LilyPad LED, as in Figure 7-19. You could go through the trouble of hooking up all your LEDs, but really, you just need to make sure that your switch components will work and that you understand the concept.

Classroom tip: Students might want to skip this step, but as with the last project, it is very important to draw out the circuits by hand. It will help your students to mentally prepare for sewing and help them to make fewer mistakes while sewing circuits. You can always tell them, "You can do it quickly and have to do it twice, or you can go slowly, and do it the right way once."

Step 4: Determine DIY Clip Switch and Circuit Route

Since this circuit route isn't as straightforward as it looks in the schematic, you'll really want to spend some time determining how your stitches will go from the "art" area of your cuff to the metal clasp DIY switch. Transfer your LEDs from your drawing to your cuff, and baste with a dab of craft glue. Wrap your cuff around your wrist as if you are going to wear it, and place the male clasp on the cuff (Figure 7-20). Baste stitch the clasp to the cuff. As you design your circuits and before you start sewing, you need to remember that you must not cross your negative and positive routes at any point; otherwise, you will short your circuit. (Although you can sometimes fix this with clear fingernail polish or hot glue, it's much easier not to make the mistake in the first place.)

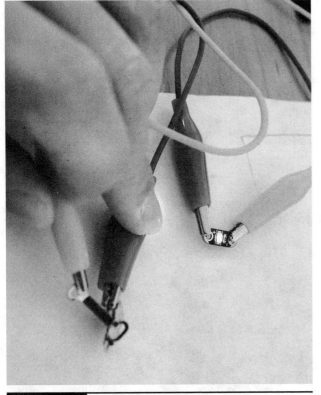

Figure 7-19 DIY switch and alligator clips test.

Figure 7-20 Placing and basting DIY switch clasp.

Step 5: Baste and Sew Battery Pack

As you did in the preceding project, you'll want to baste your battery pack with a dab of craft glue. Then you'll start by sewing the negative leads on your battery pack to the male clasp on your bracelet with your conductive thread. Look back to Figure 7-17 often to make sure that you only sew the negative pins from the battery pack to the male end of the metal clasp (Figure 7-21). Do not sew your negative pin from the battery pack to the negative pins on your LEDs or your DIY switch will not work. (This would complete the circuit, and your LEDs would remain on all the time. I did it by accident the first time I sewed this project!)

Step 6: Sew Negatives to DIY Switch Closure

Using your conductive thread, begin by sewing three to four loops on the female end of your metal clasp. Use a running stitch to sew across to the other end of the female clasp. Again, use three to four loops on this end of the clasp; then sew across to the first negative pin on the closest LED using a running stitch (Figure 7- 22).

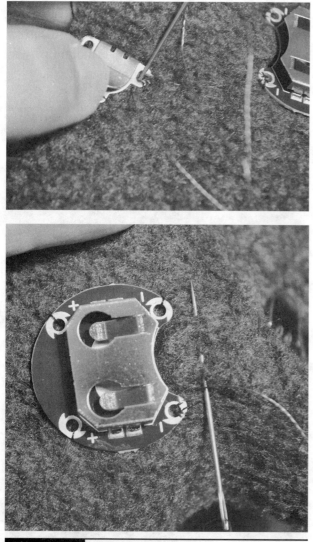

Figure 7-21   Male clasp to negatives on battery pack.

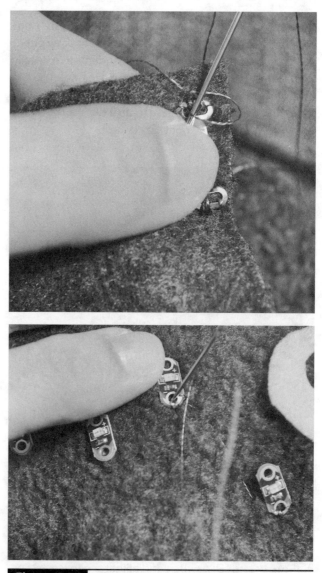

Figure 7-22   Clasp to negatives on LilyPad LEDs.

Make sure that you angle the stitches toward the bottom of your cuff so that you have extra room and will not accidentally cross your stitches when you sew your positive route in the next step.

Loop three to four times through the negative pin and continue your running stitch to the next LED. Do not cut your thread. Hold your thread tight as you did in the preceding project but not so tight that the fabric buckles. Remember that you are getting a connection through the tiny flower-like petals on the LilyPad LED. Use a basic running stitch to continue the negative grounding route to the next negative pin on the next LED. Once you have sewn from negative pin to negative pin without breaking your thread and are at the final LED, loop three or four times through that negative pin, tie a knot, and cut your thread.

Step 7: Positivity Route

We are going to sew from the LEDs back to the battery pack for our positive circuit route. Cut a length of conductive thread from one shoulder to arm's length. Thread your needle, and starting with the LED furthest from your battery pack, tie a knot in your conductive thread and secure your LED and your connection by looping through this pin three to four times.

Using a running stitch, sew across to the next LED, and loop your thread through three to four times. Without breaking your thread, continue your running stitch to the next LED. Loop through the positive pin on this LED, and remember that you should be holding your thread tight. You need tight loops around this digital pin to make sure the LED gets voltage from the battery pack. Once you've sewn to each LED, continue your positive routing to the battery pack. Because our switch is connected to

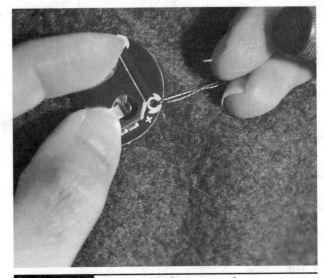

Figure 7-23 Positives on battery pack.

the negative routing, we still need to connect this line of LEDs to the positive pins on the battery pack. The battery pack is heavier than the LEDs, so you will want to loop through the first positive pin five to six times. Then continuing your stitch without breaking the thread, sew over to the other positive pin. Loop five to six times, tie a double knot, and cut your thread close to the knot (Figure 7-23).

Step 8: Test

Whew! You did it! Let's test your bracelet by connecting our DIY clasp switch. Did your LEDs light up? No? Check your routing, and make sure that your negative and positive routes do not cross. Have one LED not lighting up? Did you accidentally skip this LED in your running stitch? Or are the threads loose? Figure 7-24 shows both good and bad examples of circuitry sewing. Did you use long or short stitches? Even though it is easier to use long stitches when hand sewing, it takes away the control you have over your circuits. Instead, you should use small stitches to keep your thread from moving around.

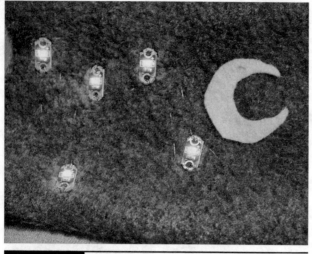

Figure 7-24 Circuit test.

Step 9: Embellish

Follow our stitch design or create your own! I used a yellow piece of felt for the moon and a chain stitch to adhere the moon-shaped felt to the cuff. I found a great technique from Sublime Stitching for an easier chain stitch at this link: http://sublimestitching.com/pages/how-to -chain-stitch.

After attaching, use a running stitch to make the moon radiate (Figure 7-25). Once the moon is glowing to your satisfaction, tie a double knot on the backside of your cuff. Use the same radiating technique on your stars; then tie off your thread.

To create the swirling nighttime sky, separate three to four threads from the blue embroidery thread and create a swirl with a running stitch. Then using white thread, stitch in between the blue stitches to create a contrasting effect.

Step 10: Wear It!

It's time to wear your newly designed wearable art cuff with pride (Figure 7-26). Make sure that you show off your DIY switch clasp and wow your friends!

Challenges

- Incorporate a LilyTwinkle, which is a preprogrammed microcontroller that has a beautiful blinking effect.

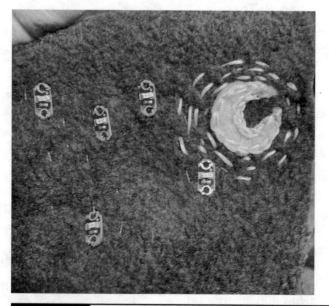
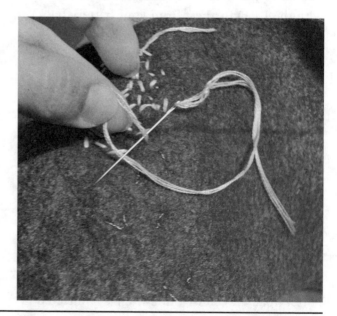

Figure 7-25   Radiating moon with running stitch.

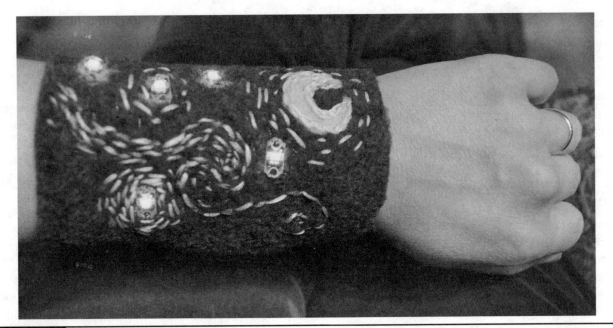

Figure 7-26 Finished wearable art cuff.

- This is a perfect project for LilyTwinkle. This would make your "Starry Night cuff" more interactive.

- What other embroidery stitches could you use to re-create a brushstroke? A seed stitch could make an interesting display!

- Could you get even more complicated and learn to program a LilyPad controller and program your lights to blink? Or even add sound effects to your bracelet?

Project 30: Heavy Metal Stuffie with Lilypad Arduino

Classroom tip: Before you get into complicated Arduino coding, have your students run multiple example sketches on the Arduino IDE using the LilyPad ProtoSnap. They can upload them directly to their LilyPad ProtoSnap by changing pins in the code as needed and immediately see the power of programming.

Cost: $$$

Make time: 2–3 hours

Supplies:

Materials	Description	Source
Fabric	Half yard of plush for guitar body Quarter yard of plush or felt for contrasting color	Craft or fabric store
Thread	Embroidery thread (matching for guitar accents, contrast for outer edging, and silver for strings)	Craft or fabric store
Arduino Tone Tutorial	Simon Monk's Tone Tutorial	www.arduino.cc/en/ Tutorial/ PlayMelody
Full kit	LilyPad ProtoSnap Kit (kit includes items listed below), SparkFun (DEV-1126)	SparkFun
LilyPad Simple	LilyPad Simple, SparkFun (DEV-10274)	SparkFun
LilyPad Vibeboard	LilyPad Vibeboard, SparkFun (DEV-11008)	SparkFun
LilyPad Buzzer	LilyPad Buzzer, SparkFun (DEV-08463)	SparkFun

(continued on next page)

Materials	Description	Source
LilyPad Light Sensor	LilyPad Light Sensor, SparkFun (DEV-08464)	SparkFun
Conductive thread	Conductive thread, SparkFun (DEV-10867)	SparkFun
Micro-USB adapter	LilyPad FTDI Basic Breakout: 5 V, SparkFun (DEV-10275)	SparkFun
Battery	Lipo battery, SparkFun (PRT- 00731)	SparkFun

Classroom tip: Remember to stay close by if working this project with students. This is an advanced project for students who've sewn circuits before. Take note that SparkFun does have a warning on this kit and battery: "Please use caution when using this battery in wearable projects. When using conductive thread, a short in the thread can create sparks and heat. We recommend using coin cell batteries for beginners."

Step 1: Arduino Madness

If you haven't installed Arduino software onto your computer, you need to do that first. If you have, you'll still need to download the FTDI driver so that you can upload code to your LilyPad. There is a great tutorial at SparkFun: www.sparkfun.com/tutorials/308. This will help you to download the needed software and explain how to run through a few example sketches.

How Does the ProtoSnap Work?

The LilyPad ProtoSnap is already wired up, so you can run an example Arduino sketch and see it running without having to wire anything. The current model has hidden pins for some of the components, so play a lot before you decide to snap it apart because you'll have to rename the pins once you start writing your own sketches.

Classroom tip: Make sure that all student computers are preloaded with Arduino software and the FTDI driver. Follow the download instructions from SparkFun: www.sparkfun.com/tutorials/308. With the ProtoSnap attached to student computers via the FTDI, run two to three example sketches from each "Built-in example" by going to File > Examples and then exploring sketches. Take the time to point out how the code works. For example, in the "Blink" sketch, point out how pin 13 is defined in "Void Setup." Then have students attempt to change this pin to a different LED on their ProtoSnap. Note that they must also change pin 13 in the "void loop" or nothing will happen on the ProtoSnap. In the "Blink" sketch, they need to change pin 13 in the digitalWrite area of the code if they want to light up the correct LED. Can they add another LED by adding a second pinMode and digitalWrite? Lastly, challenge them to change the delay and see what happens to their program. Repeat this type of tinkering with sketches to find out more about how the LilyPad ProtoSnap is programmed with Arduino.

Step 2: Run Example Sketches

Run a few example sketches on the ProtoSnap, and tinker with some of the code in the sketches to get some basic ideas of how Arduino sketches work on this microcontroller and its components. When you are ready to play music, you are going to unsnap your ProtoSnap because we are going to rename pins when we upload this "Iron Man" code to the board.

Step 3: Upload "Iron Man" Code to the Board

```
/*
Big Book of Makerspace Projects - Project 21 Heavy Metal Stuffie with Iron Man Tones
Written with the help of Denton Public Librarian Trey Ford and based on Simon Monk's
tone tutorial. For more info on tone go to: http://arudino.cc/en/Tutorial/Tone
*/
//Light Sensor connections:
// S pin to A5
// + pin to +
// - to -

// Transistor (if flat head is pointing up- if sewing on back of fabric, remember to
// switch right leg and left leg!)
// A3 to 330 ohm to middle pin
// Right leg to -
// Left leg to PIezo -

// Piezo(Buzzer) connections:
// + to +
// - to -

// Vibeboard connections:
// + to A2
// - to -

/*************************************************
 * Public Constants
 *************************************************/

//This section is the notes for Iron Man

#define NOTE_C4 262
#define NOTE_CS4 277
#define NOTE_D4 294
#define NOTE_DS4 311
#define NOTE_E4 330
#define NOTE_F4 349
#define NOTE_FS4 370
#define NOTE_G4 392
#define NOTE_GS4 415
#define NOTE_A4 440
#define NOTE_AS4 466
#define NOTE_B4 494
#define NOTE_C5 523

int sensor = A5;
int ledPin = 13;
int vibeBoard = A2;
int lastnote = 0;
```

```
int piezo = A3;
int melody[] = {
NOTE_E4, NOTE_G4, 0, NOTE_G4, NOTE_A4, 0, NOTE_A4, 0, NOTE_C5, NOTE_B4, NOTE_C5,
NOTE_B4, NOTE_C5, NOTE_B4, NOTE_C5, NOTE_B4,NOTE_G4, NOTE_A4, NOTE_A4
};

// note durations: 4 = quarter note, 8 = eighth note, etc.:
int noteDurations[] = {
2, 2, 8, 4, 4,16, 2, 8, 8, 8, 8, 8, 8, 8, 8, 4, 2,2, 2
};

void setup() {

pinMode(ledPin, OUTPUT); // This set of instructions,set the the ledPin
                         // and vibeBoard to be outputs
pinMode(vibeBoard, OUTPUT);

}
void loop(){
digitalWrite(ledPin, LOW); //Turns off the LEDs for normal operation
digitalWrite(vibeBoard, LOW);

if(analogRead(sensor) < 5) // If you strum the guitar (or cover the sensor)
                           // then the piezo will go off
{
digitalWrite(vibeBoard, HIGH);

// to calculate the note duration, take one second
// divided by the note type.
//e.g. quarter note = 1000 / 4, eighth note = 1000/8, etc.
int noteDuration = 1000 / noteDurations[lastnote];
tone(piezo, melody[lastnote], noteDuration);

// to distinguish the notes, set a minimum time between them.
// the note's duration + 30% seems to work well:
int pauseBetweenNotes = noteDuration * 1.30;
delay(pauseBetweenNotes);
// stop the tone playing:
noTone(piezo);
digitalWrite(ledPin, HIGH);
lastnote++;

}
if(lastnote >= 19){ //Resets song if last note is at the end
lastnote = 0;
delay(100);
}
}
```

Step 4: Snap ProtoSnap Pieces

If you'd like to program a different song, now is the time to test it out. You can keep most of the code the same and only change the melody and the note lengths to play your own favorite song. If you want to keep "Iron Man," then snap everything apart with wire cutters and clean edges as in Figure 7-27. Hook up alligator clips to the pins specified in Table 7-1, upload your code, and test. If it doesn't work, check to make sure that you've clipped alligator clips to the right pins and copied the code directly. (You can copy the code from download files available at www.mhprofessional.com/.)

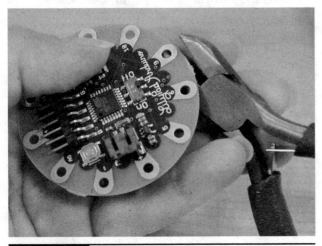

Figure 7-27 Snap ProtoSnap.

TABLE 7-1 LilyPad Setup

Light Sensor	Transistor (Flat Head Up)	Piezo (Buzzer)	Vibeboard
S pin to A5	A3 to 330 ohm to middle pin	+ to +	+ to A2
+ pin to +	Right leg to –	– to –	– to –
– to –	Left leg to piezo –		

Step 5: Cut Guitar Template

Locate the guitar template in the back of the book and cut out the template. Fold your plush in half, and pin template to your fabric with the base of the guitar flush on the fold.

Step 6: Cut Guitar Out of Plush

Use the template to cut a guitar shape out of plush. You can design your own guitar shape here, but remember to trace your template on the fold of the fabric so that you cut two shapes. Also, make sure that you cut on the fold (but don't cut the fold) so that your shapes are already connected (Figure 7-28).

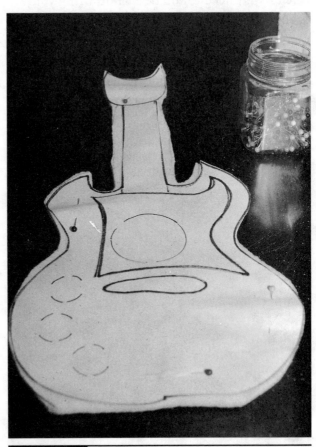

Figure 7-28 Cut on fold.

Step 7: Cut Accents from Template with Pocket for Battery

Using the template and your contrasting fabric, cut the guitar accents and a small pocket for the Lipo battery. Pin these accents to the guitar, and layer the battery pocket on top of the body accent.

Step 8: Stitch Accents to Main Guitar

You will use a fell stitch to join the accents to the main body of the guitar. This stitch is also known as an *applique stitch*, and we are using it because it is strong and has a nice look. You'll be working from right to the left. Cut a piece of embroidery thread about a shoulder's length, thread your needle, and tie a knot. Bring your needle up from under the accent but above the main fabric. This will hide your tail end inside the accent. Insert the needle above and parallel to where your needle came out, and then angle it down toward where you want the next stitch to come out, as in Figure 7-29. Continue until

you've attached the accent, battery pocket (make sure that you sew only around the pocket and you don't accidentally close it up), and guitar head accent.

Step 9: Map Circuitry and Lay Out Components

Follow circuitry templates and lay out components on your project using Figure 7-30 as your guide. Baste components with glue, taking note to match + on components to the + on the templates. As with the last project, you do not want to accidentally cross any of your circuit routes because you want the current to flow in a closed loop through each component, and crossing would cause a short circuit.

Step 10: Test with Alligator Clips

Yes, we already tested your wiring, but it doesn't hurt to test one more time before sewing! Check

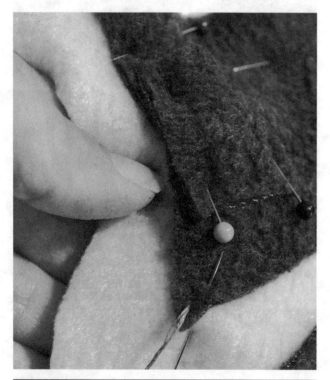

Figure 7-29 Sewing accents.

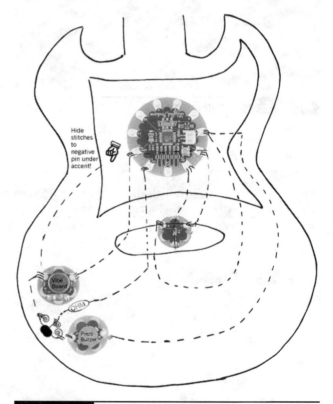

Figure 7-30 Guitar circuit routes.

to make sure that you have your transistor hooked up to make the piezo as loud as possible. Flip your transistor, and try again. Which way is loudest? Remember that you will sew this on the inside of your guitar, and you do not want to have to sew this twice. So take note of which direction is loudest, and write down how you will need to place this transistor to make your guitar/piezo as loud as possible.

Step 11: Prepare Transistor and Resistor

Using a pair of needle-nose pliers, grab the end of a leg on your transistor and curl it around the tip of the pliers so that it makes a spiral (Figure 7-31). Repeat for each leg. Do the same with your 330-ohm resistor. This will make it easier to sew these components on and get a good connection with your conductive thread.

Figure 7-31   Use needle-nose pliers to curl component legs.

Step 12: Sew Transistor to Resistor

On the wrong side of the fabric, dab a little fabric or craft glue to the transistor and resistor to baste them to the fabric. You will be sewing these inside the guitar body so you can safely insulate them in the last step with felt. Make sure that you put the transistor flat side down on the fabric (unless when you tested your transistor and it was louder the other way). Thread conductive thread onto needle, tie a knot, and loop five to six times around the middle pin on the transistor. Make sure that you stitch your thread very close around the wire because this is how the current will flow through these components. Stitch along the individual wire and not the whole spiral; once your connection is secure, hide a running stitch up to the resistor. Repeat the process for sewing to the spiraled leg of the resistor, and tie off. Use stretchy fabric glue to insulate your connections. Refer to Figure 7-32.

Cut another shoulder's length of conductive thread, tie a knot at the base, and loop three to four stitches around the other leg of the resistor. Remember to keep your tail end on

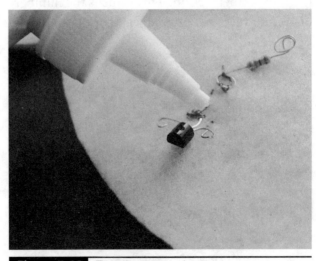

Figure 7-32   Transistor to resistor.

this side of the fabric because we are sewing the resistor to the wrong side of the fabric. Hide a running stitch up to pin A3 of the LilyPad by going through the center of the fabric and not letting your needle come up on the right side of the fabric. Figure 7-33 shows this technique. Bring the needle to the right side of the fabric through pin A3. Stitch tight loops around pin A3. This is the pin that's going to tell your piezo (by going through the resistor and transistor first) the guitar melody you want it to "play." You could sew this project without a transistor, but the piezo would be awfully quiet. The transistor turns up the volume on the piezo by taking the small input current and allowing more current to flow to the piezo, which, in turn, makes the piezo louder. When the buzzer "plays" at a certain frequency, it creates a tone. Each identified frequency plays a note or a pitch. We can control that pitch by entering the hertz value of the given tone we want our buzzer to play. (If you haven't, go check out more about the science of sound in Chapter 6!) You will notice at the beginning of our code (and in Table 7-2) that each note is identified by its hertz value. If you want to play a different song, you'll have to make sure that you define the note in this section! This value is the number of sound waves per second whether the sound

waves are being made by a buzzer or a flip-flip on a PVC pipe, as you discovered in Chapter 6. You learned more about hertz and sound waves in Chapter 6, so go back and revisit that chapter if you are unclear about hertz and sound waves. In Arduino coding, you can also control the length or *duration* of a note, as you learned in Chapter 5. We are going to set that duration in a collection of variables known in the Arduino world as an *array*. See Table 7-3 for the array of note durations so that if you want to adjust the length of notes, you will know where to adjust your code.

TABLE 7-2   "Iron Man" Tones

Note Value	Hertz Value
#define NOTE_C4	262
#define NOTE_CS4	277
#define NOTE_D4	294
#define NOTE_DS4	311
#define NOTE_E4	330
#define NOTE_F4	349
#define NOTE_FS4	370
#define NOTE_G4	392
#define NOTE_GS4	415
#define NOTE_A4	440
#define NOTE_AS4	466
#define NOTE_B4	494
#define NOTE_C5	523

TABLE 7-3   Note Duration Array

```
// note durations: 4 = quarter note,
// 8 = eighth note, etc.:
int noteDurations[ ] = {
2, 2, 8, 4, 4,16, 2, 8, 8, 8, 8, 8,
8, 8, 8, 4, 2,2, 2
};
```

Step 13: Sew Positive Vibes

Starting with a new thread, tie a knot and loop three to four times around the positive pin on the Vibeboard (Figure 7-34). Sew a visible running stitch toward the sensor, as in Figure

Figure 7-33   Sew to pin A3.

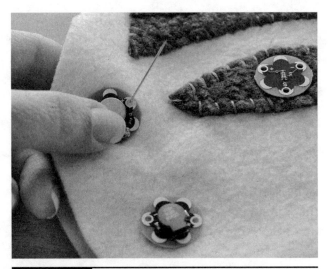

Figure 7-34 Positive pin.

Figure 7-35 Running stitch.

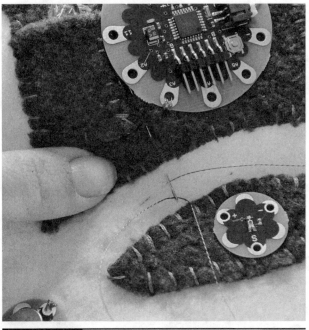

Figure 7-36 Sew to pin A2 and tie off.

7-35. Then turn your stitching toward pin A2 on the LilyPad, as in Figure 7-36. Sew up pin A2, loop four to five times, and tie a secure knot on the wrong side of the fabric. Cut your thread. This route will send the power to your Vibeboard so that when activated by the light sensor, your guitar will shake a little in your air guitar player's hands. Of course, it won't work until you sew your negative circuit and complete all circuit routes.

Step 14: Sew Positive Piezo

With a new thread, tie a knot, and attach the piezo by sewing three to four loops around the positive pin on the piezo. Bring your running stitch around the base of the guitar, as in Figure 7-37. Then hide your thread alongside the body accent, and sew across to the + pin on LilyPad. Loop three to four times around the positive pin, and tie off on the wrong side of the fabric.

Step 15: Sensor S to Pin A5

Use a needle to poke through to the wrong side of the fabric, and mark with a sharpie where pin A5 is located on the wrong side of the fabric. Because our circuit traces are growing, we need to make sure that we sew directly from the S pin on the sensor to LilyPad pin A5, as in Figure 7-38. This trace is going to give the directions to the sensor to play the piezo buzzer only when the light is low, that is, when you play air guitar! Make sure that you insulate the knots on the back of the fabric with stretchy fabric glue.

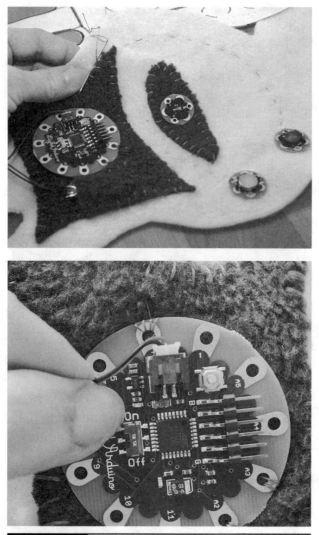

Figure 7-37   Piezo to +.

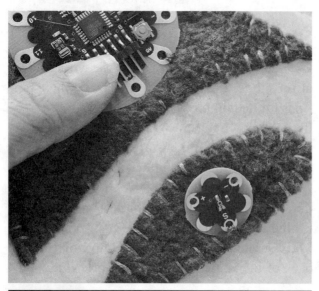

Figure 7-38   Sensor to pin A5.

Step 16: Sensor + to + on LilyPad

Because the S pin signals the sensor, we still have to send power to the sensor so that it knows to take a reading. To do this, you will sew a circuit from the + pin on the sensor to the + pin on the LilyPad. You have already sewn to this pin, but it is okay to branch your circuit route by sewing to this pin again. Just as you did in previous steps, starting with a new thread, loop through the + pin on the sensor and sew a running stitch up to the + pin on the LilyPad. Loop through this pin four to five times, and tie off. Use fabric glue to insulate on the backside of the fabric. Remember to follow the routing outlined in the guitar template (Figure 7-30).

Step 17: Sensor – to – on LilyPad

Ground the sensor, and complete the circuit by sewing three to four loops around the – pin on the sensor, and then sew a running stitch to the – pin on the LilyPad.

Step 18: Transistor to Vibeboard to – on LilyPad

On the backside of the fabric, loop around the right leg of the transistor three to four times, and bring your needle to the front side of the fabric. Make a visible running stitch to the negative pin on the Vibeboard, as in Figure 7-39. Continue running your stitch up by the battery, and then take your needle to the backside of the fabric, hide your stitch as in Figure 7-39, and bring your needle back to the front by coming out on the negative pin on the LilyPad Simple. Loop three to four times around the – LilyPad pin and knot and tie off on the backside of the fabric.

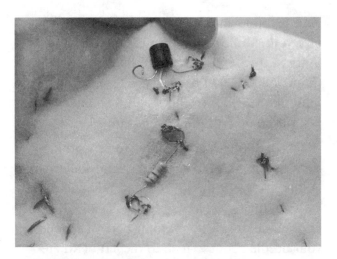

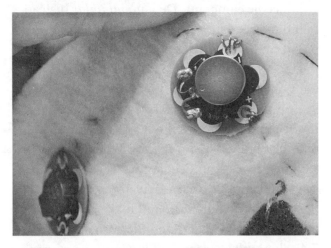

Figure 7-39   Negative route.

Step 19: Piezo to Transistor

Starting at the – pin on the piezo with a freshly cut thread, loop through this pin three to four times, and sew a running stitch up to the closest transistor leg and loop three to four times around the leg; then knot and tie off. (Figure 7-40).

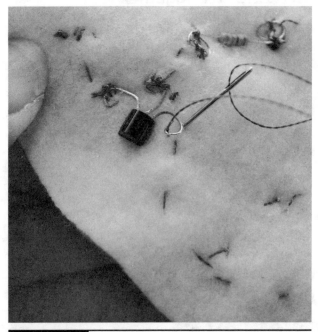

Figure 7-40   Connecting the piezo to the transistor.

Step 20: Check Your Circuitry and Test

Make sure that you've sewn from every component to the LilyPad Simple board as directed in the template and Figure 7-30. Turn on your guitar, and test! Did it work? Hooray! Move to the next step. If it didn't work, make sure that you have tight loops around the pins and that you've sewn the correct circuit routes. Also look back over your threads and make sure that you haven't crossed any lines. If you have, you can use fabric glue to insulate your stitches or add a piece of felt under a stitch, but this is not ideal. If it still doesn't work, you may have circuits crossed in the fabric, and you'll need to identify where those are and start those stitches over.

Step 21: Insulate Threads

Use the original template to trace the guitar shape on interfacing or adhesive felt. Then measure ½ inch in from the edge, and mark on this backing. Cut and place over circuits as in Figure 7-41.

Step 22: Stitch Together

Place the guitar sides together, and starting from the bottom near the fold, pin the sides together to keep the edges together while you sew. Using a blanket stitch and contrasting thread, stitch up guitar. The most important thing is that you keep your stitches the same distance apart

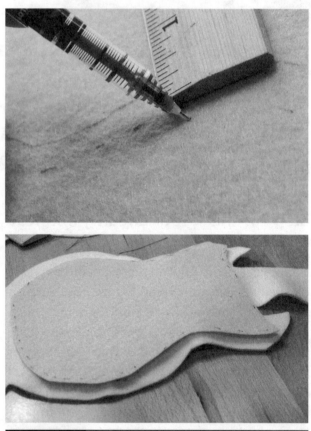

Figure 7-41 Insulating threads with adhesive felt.

and the same distance from the edge. This will give your stuffie a more professional look. Start by threading silver embroidery thread (nonconductive) onto your needle and tying a knot on the end. For this stitch, I actually found it easier to tie both thread ends together.

For this stitch, we are going to be working right to left. So, starting at the right side of your guitar, take your threaded needle and come between the two layers so that you can hide the knotted tail inside your guitar body. Then insert your needle about ⅛ inch to the left of where your needle came out. *Do not pull tight!* Instead, pull the thread most of the way, but before you pull it tight, bring your needle back up through the loop you just created. As you pull the thread tight, make sure that the floss forms a straight line on top of your two fabric pieces. Repeat this technique across the edge of the guitar, and refer to Figure 7-42 for details. When you get to a corner, place your needle back through your last stitch, and as you pull the thread through the loop, turn your work and make sure that the thread makes a diagonal to the corner. Figure 7-43 shows the corner details. Continue with blanket stitch as before until you get three-quarters of the way around the neck of your guitar.

At the neck, stuff before you continue your blanket stitch because this would be difficult later on. After stuffing, continue the blanket stitch until you are about 2 inches away from the fold. Do not cut your thread or tie off. Instead, leave thread in the needle, and pin to the side so that you can tie off *after* you stuff your guitar (Figure 7-44).

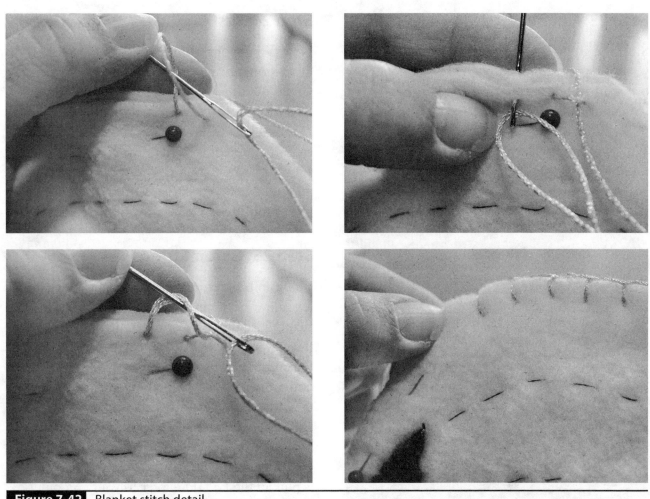

Figure 7-42 Blanket stitch detail.

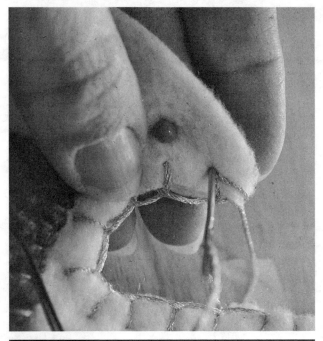

Figure 7-43 Corner details.

Figure 7-44 Stuff neck.

Step 23: Get Stuffed

Stuff the guitar with stuffing. Make sure that you push the stuffing into corners, and don't overstuff or understuff. When you are happy with your stuffie, pick up the needle you pinned to the side, and finish sewing your guitar shut. When you reach the end, tie off and hide the tail knot inside of the guitar (Figure 7-45).

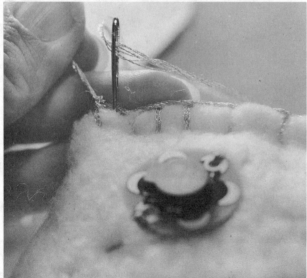

 **Figure 7-45**   Sewing together.

Step 24: Test

Now is the moment of truth. Does it work? Turn it on, and cover the sensor. Does it work? Congratulations! No? Do some troubleshooting. Did you sew all the correct circuit routes? Double-check your routing, and make sure that each component is sewn to the correct pin on the LilyPad Arduino board. Are your stitches tight and neat? Did you sew close to the pins? Are any routes touching and shorting out circuits?

Step 25: Guitar Strings

Using embroidery thread, tie a knot under the felt guitar head, and use a loose running stitch to mimic a guitar string down the neck and to the guitar body. When you get to the accent on the body, tie your end knot, and hide the tail under the felt accent (Figure 7-46). You can use this time to tie down your battery wires from the Lipo battery (Figure 7-47). You shouldn't have to remove it even when the charge is down because all you need to do to charge this battery is plug in the FTDI adapter to your computer.

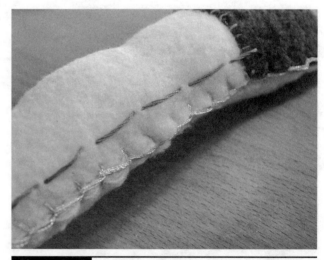

Figure 7-46   Adding strings.

Figure 7-47 String tips and securing battery wire.

Step 26: Play Air Guitar!

Play air guitar, and when you get tired of this song, reprogram it (Figure 7-48). All you have to do is figure out the notes of your favorite songs, define them in the code, and adjust the length in the note duration section.

Figure 7-48 Air guitar!

Challenges

- Can you lengthen the array and include more notes from the song "Iron Man"?

- Can you reprogram the guitar to play one of your favorite songs?

- Need more bling? Add LEDs in a parallel circuit up the guitar neck.

- Or even learn how to fade the RGB LED, and add another "knob" on your guitar. Remember to not cross circuit routing, or if you need to, you can cover one route with felt and add your other circuit route over the top, as long as the conductive stitches do not touch!

"E-Textile" Challenge

Can you design your own stuffie that does something? What will you program it to do? Adjust an example sketch to do your bidding, and make your own e-textile project.

Make sure that you take pictures of your challenge project. Tweet it to us @gravescolleen or @gravesdotaaron or tag us on Instagram and include the hashtag #bigmakerbook to share your awesome creations. We will host a community gallery of projects on our webpage.

Makey Makey

HERE ARE SOME quick and easy project ideas for Makey Makey. The Makey Makey is an invention kit designed and created by Eric Rosenbaum and Jay Silver (Figure 8-1). This microcontroller plugs into your computer via USB and makes your computer think you are hooking up an external keyboard. The Makey Makey has six inputs on the front that allow you to control your arrow keys, spacebar, and the click of your mouse. On the back, there are even more keys you can program. Try hooking up real bananas to your arrow keys by attaching an alligator clip from a banana to an arrow input on the Makey Makey. Then attach another alligator clip to one of the inputs on the bottom row labeled "earth." Now, if you hold the metal on the alligator clip connected to earth and then you touch the banana with your other hand, you will complete a circuit and trigger the key that the banana is hooked to!

Go to makeymakey.com/piano and plug all your alligator test leads into different bananas, and now you can play the banana piano! You don't have to just play piano; you can actually control any webpage with bananas. You can control a slide deck, a video game, anything you put your mind to! Since you know how to create your games in Scratch now, you can easily create your own custom game controllers with themes that match your games! (Hello 1980s video arcade!) When you're ready to get beyond the banana, we recommend going through the simple circuit challenge available at http://makeymakey.com/lessons/simple-circuit -challenge/.

Get ready to get inventive with these projects that will help you to start inventing your own switches and give you infinite possibilities for invention with your Makey Makey.

Classroom tip: Teachers, did you know that there is a whole suite of lessons for Makey Makey? Check them out at makeymakey.com/lessons!

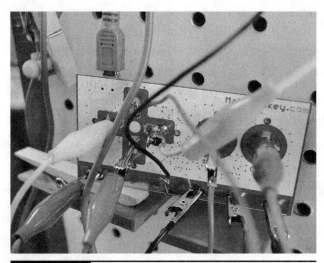

Figure 8-1　Makey Makey invention kit.

Chapter 8 Challenge

"Makey Makey assistive technology" challenge.

Projects 31–33: Building Swing Switches, Fold Switches, and Pressure Sensors

Makey Makey is a great way to make your space interactive. You can make your own sound switches, camera triggers, and timers. Projects 31 through 33 will give you some great ideas on ways to invent switches for Makey Makey. One of the best things about Makey Makey is that you can think of the world as your invention kit! You can even start by looking in your kitchen!

Project 31: Build a Swing Switch

One of our favorite switches for Makey Makey is the swing switch. It's especially great for use with moving objects if you want to trigger a timer or sound effect.

Cost: Free–$

Make time: 10–15 minutes

Supplies:

Materials	Description	Source
Invention kit	Makey Makey, alligator test leads, USB cable and wires	JoyLabz
Everyday stuff	Tape, straws, skewers or chopsticks, aluminum foil, and cardstock paper	Office supply store Grocery store
Tools	Scissors, box cutter	Craft store
Base supplies	Cardboard, Legos, Duplo blocks, or wooden blocks	Recycling Toy box

Step 1: Assemble a Swing Switch

The switch we are constructing works just like a swing on a swing set. In this switch, an object such as a toy car or train will push a foil swing, and it will run into another conductive surface to complete a circuit and trigger a key on your computer. Start by constructing two posts that are about 2 inches taller than the object that is going to push the swing. If you are using a toy with a track to trigger the switch, have the track and toy in place to get an idea of how high and far apart the posts will need to be. Directly behind the first set of posts, add a second set, about 1 inch lower than the first. If you are using Duplo bricks such as those in Figure 8-2, usually one brick lower works perfectly.

Step 2: Beams and Straws

Skewers make a great beam for swing switches to rotate around. Lay a skewer across the two posts, and use a pair of scissors or wire cutters to mark it. You don't have to cut all the way through it; just bend it back and forth until it breaks. Cut two straws the width of the post, as in Figure 8-3. Tape these straws in place at the tops of the

Figure 8-2 Swing posts and skewer beams.

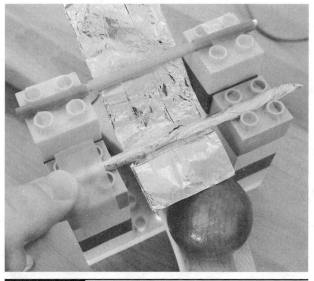

Figure 8-4 Weighted bottom and placement of swing.

Figure 8-3 Straw center.

posts. The straws will help to keep your skewer cross-beam in place when the swing rotates.

Step 3: Construct a Swing

Cut a straw that is just slightly smaller than the width between the two posts (see Figure 8-3).

Your swing will be different, but for this one we used an 8- × 8-inch sheet of foil folded into an 8- × 2-inch strip. It's important to weight the bottom end of the swing so that it will return to its resting position and not stay activating the switch. To do so, simply fold the strip over in a couple of 1-inch folds so that you now have a six strip, as in Figure 8-4.

Add the straw onto the cross-beam, and hold the strip in place. Center the strip, and ensure that it has ample clearance at the bottom. Tape it in place, and test it by swinging it back and forth. Add weight to the bottom if needed; however, try to keep it as light as possible so that you don't slow down the object that is triggering the switch.

Step 4: Clothesline

The next part of this switch is the wrestling equivalent of performing a clothesline move but with a swing. This stops the swing from moving forward and completes the circuit, which will trigger your Makey Makey. Our "clothesline" is going to rest on the second set of posts just slightly lower than the first set so that as the foil swing moves, it will make contact with the clothesline, as in Figure 8-4. Wrap the remaining skewer up with foil, and tape it on top of the second set of posts. Most likely you will have to adjust the position of the skewer by performing a few test runs until you have the perfect height and distance from the swing. If it is too low, it will stop the trigger object, and if it is too high, it may not trigger the Makey Makey.

Step 5: Swing, Set, Match Switch

Connect the Makey Makey to your computer, and then clamp an alligator clip to the spacebar or desired arrow key input. Connect another alligator clip (also known as a *test lead*) to earth on the Makey Makey and the other end of the lead to the "clothesline" beam (Figure 8-5). Take a couple of tests runs to see if your object will trigger the Makey Makey switch, and make any

necessary adjustments. If the light on the key you've clipped to on the Makey Makey comes on when the foil connects, then you've made a successful switch (Figure 8-6).

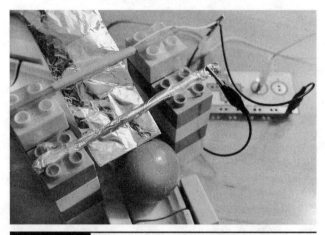

Figure 8-6 Test.

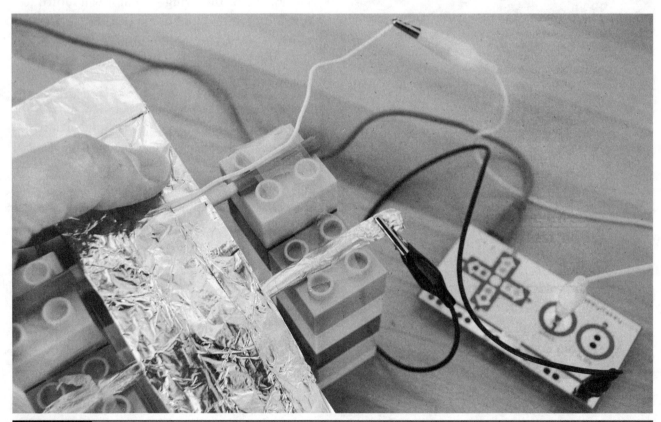

Figure 8-5 Hookup.

Challenges

- What other tools could you use to make a swing switch?

- Could you make one that functions more like a door?

- What about using littleBits to make an automated swing switch?

Classroom tip: When using Makey Makey, the only thing holding you back is your imagination! For students, I recommend gathering a very large box of bricolage or random junk and making an inventor's box. Keep lots of different types of conductive and nonconductive objects in the box and make testing conductivity part of the fun.

Project 32: Building a Fold Switch

The fold switch is very versatile, and of all the switches, it is probably the easiest to make. Another of its strengths is that it can be made small enough for a toy train to roll over or large enough to be used to create a talking book drop.

Cost: Free–$

Make time: 10–15 minutes

Supplies:

Materials	Description	Source
Invention kit	Makey Makey, alligator test leads, USB cable, and wires	JoyLabz
Nonconductive building supplies	Cardstock paper, masking tape or clear tape, cardboard	Craft store / Recycling bin
Conductive stuff	Aluminum foil	Grocery store
Tools	Scissors, box cutter	Craft store

Step 1: Construct a Fold Switch

This switch can be made with a variety of papers, but it is important for you to consider how much pressure will be applied. A switch that will pressed by a finger could be made with light paper, but a switch that will be used for a book drop or stepped on might require a material such as poster board, cardboard, or even rubber floor mats!

Once you decide on your material, start by folding it in half horizontally. If you decide to use a long, narrow piece of regular cardboard, you'll want to use a box cutter to score one side of the center of the piece of cardboard so that you can fold it. Remember, do not cut all the way through. Snap the cardboard along the cut to create a crease, and fold it over.

Step 2: Tape and Foil

Cut two pieces of foil about 2 inches longer than the paper or piece of cardboard so that you can attach your leads on the end of the switch. Fold the strips over to create a 1- to 2-inch strip of foil. Center the strip, and line one end of the foil strip up with the right edge of the switch. You should have a flap hanging over the edge that we will clip our Makey Makey to in the next step. Use tape to secure the foil strip to the cardboard (Figure 8-7). Turn the base around and repeat the process, except this time line up the end of the foil with the left side. You should now have an aluminum flap on each side that you can fold over and place your test lead or alligator clip (Figure 8-8). You can also use hookup wire or strip old telephone wire and tape it directly to your foil. If you dislike the bulkiness of the head on the alligator clip, these other types of wires are preferable. There are instructions on stripping telephone wire at http://makeymakey .com/lessons/simple-circuit-challenge/.

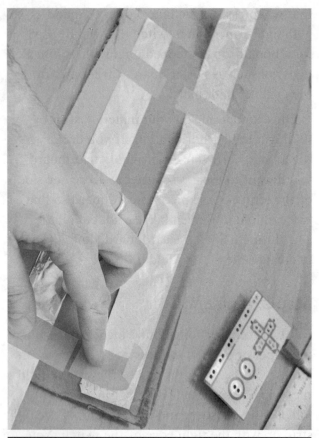

Figure 8-7 Taping foil strips.

Figure 8-8 Making loops and taping.

Step 3: Connect

Fold the flap over a few times if you need to make a stiffer area to which to connect an alligator clip. Attach one alligator clip test lead to the earth on the Makey Makey and the other end to your newly created switch. Grab another alligator clip, and clip one end to the down arrow location on the Makey Makey and the other end to the opposite end of your switch. Press down on the switch to test it! If the green light on the down arrow lights up as in Figure 8-9, then you have made a successful fold switch! If it doesn't light up when you press on the cardboard, check the placement of your aluminum. Check to see if your alligator heads are clipped to the foil because if they are connected to just cardboard, your switch will not go off. Alternatively, if your switch is constantly activated, you will need to place a buffer between your two pieces of cardboard (look at the pressure-sensor switch in Project 33). If you put both leads on one end of the cardboard, you would also constantly activate your switch.

Challenges

- Could you make this switch with something other than tinfoil?
- In what other ways could you make a switch like this for people to interact with?
- Could you make a floor mat switch?

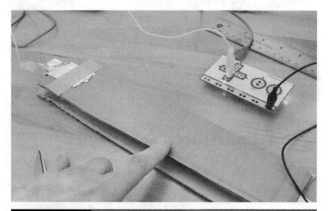

Figure 8-9 Press the fold switch.

Project 33: Building a Pressure-Sensor Switch

The pressure sensor is one of the most commonly used switches with Makey Makey. It's a great way to delight and mystify your friends!

Cost: Free–$

Make time: 10–15 minutes

Supplies:

Materials	Description	Source
Invention kit	Makey Makey, alligator test leads, USB cable, and wires	JoyLabz
Nonconductive supplies	Cardstock paper, masking tape or clear tape, cardboard	Craft store / Recycling bin
Conductive stuff	Aluminum foil	Grocery store
Tools	Scissors, box cutter	Craft store

Step 1: Make a Pressure Sensor

Cut a regular piece of cardstock in half. If you are using standard letter-size paper that is 8½ × 11 inches, you will now have a 5½- × 8½-inch rectangle. We are going to cut out a spot on this piece of paper for our Makey Makey to make a connection. So take just one piece of your freshly cut paper, fold it in half, and remove a rectangle from the center of the folded side as in Figure 8-10. When you remove the rectangle, it

Figure 8-10 Cut out rectangle.

should give you a larger and longer rectangular window when you unfold the paper.

Step 2: Under Pressure

Cut a piece of foil about 1 inch wider than the hole but long enough to stick out of one side of the paper. Use tape to attach the foil to the paper, and leave the foil hanging over the edge for now (Figure 8-11). Turn the paper over so that you can only see the window with the foil behind.

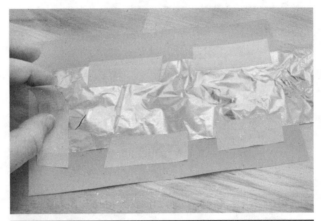

Figure 8-11 Taping foil behind the window.

Step 3: The Other Side of Earth

Cut another rectangular piece of foil 7½ × 5½ inches so that it will just fit inside the other piece of paper on three edges, as in Figure 8-12. Leave one end of the foil overhanging about ½ inch.

Figure 8-12 Foil piece for top (*left*) and bottom (*right*).

This is where we will connect our arrow key lead to the Makey Makey.

Using masking tape or painter's tape, tape the rectangle very close to the edge on three sides, as in Figure 8-13. Do not tape the inside edge because this will give a ½-inch overhang on one side and a space for you to attach your Makey Makey. Trim some of the overhang away to create a tab to which you'll attach a test lead, as in Figure 8-13.

Figure 8-13   Taping top and test lead tab.

Step 4: Give Me Some Space!

This switch sometimes works if you just lay the two pieces directly over each other. The paper window acts as a spacer until someone presses on it to complete the circuit. You can vary the width of the paper to act as a spacer, or you can add a slight bow to the top piece of paper. Bend the paper slightly, and then tape the right side of the top piece of paper about $\frac{1}{16}$ to $\frac{1}{8}$ inch away the right edge of the bottom piece of paper. Do the same for the left side, and this should give your switch a slight upward bow. You can adjust the bow later by moving one side further away or closer to the edge. Your switch should now look like Figure 8-14.

Figure 8-14   Creating a bow.

Step 5: Connect and Apply Pressure

One of the brilliant things about this switch is that you have a flat surface on top to label or decorate. We added a left arrow because that is the key we want to trigger with this switch. Attach one alligator clip to the foil tab on the top piece of paper, and connect the other end to the left arrow on the Makey Makey. Connect another alligator clip to the foil tab on the bottom piece of paper, and connect the other end to earth on the Makey Makey. Then put some pressure on your switch! Does your left arrow light up? Awesome! No? Double-check that your test leads are connected and that your Makey Makey is plugged into the computer. (You have no idea how many times we've forgotten to hook into the computer!) If your switch is constantly activated, add more space or make a smaller hole for your connection spot. You only want this switch to work when someone presses on it (Figure 8-15). This connection is made when the two pieces of foil connect. If they are connecting all the time, then you haven't made an on/off switch!

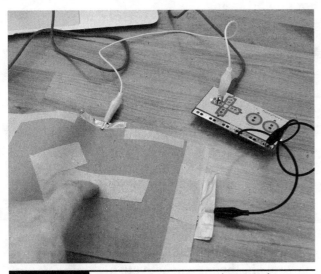

Figure 8-15 Applying pressure to the switch.

Challenges

- You don't have to use paper; we've seen many varieties of pressure switches with the Makey Makey from rubber floor mats to baking sheets and more!

- What can you do with these pressure switches? Create a candid camera? A "selfie machine"? Maybe create an interactive "moonwalk" that plays "Billie Jean" as you dance?

- Can you create cardboard bongos? A giant Pacman controller made of Pacman drawings?

- Could you make a "mac and cheese timer" out of mac and cheese?

Classroom tip: Whatever materials you decide to use, you may want to pre-cut equal strips of foil, cardstock, etc. However, if a student wants to vary the size, let him or her! Makey Makey is all about making your own unique inventions!

Project 34: Makey Makey Paper Circuit

Now that you have made a few switches, let's make something a little more complicated and even more fun! This paper circuit project is almost like making your own midi synthesizer, but out of paper. This is a superfun way to program a collage or a piece of paper with music samples to make all your DJ dreams come true! Make on!

Cost: Free–$

Make time: 30–45 minutes

Supplies:

Materials	Description	Source
Invention kit	Makey Makey, alligator test leads, USB cable, and wires	JoyLabz
Conductive stuff	Copper tape or soft conductive tape in Makey Makey Inventor's Booster Kit	SparkFun JoyLabz
Nonconductive building supplies	Cardstock paper, masking tape, or clear tape	Craft store Recycling bin
Images	Magazine cutouts and drawings	Recycling bin
Tools	Scissors, box cutter	Craft store

Step 1: Get Grounded

First things first: you need to lay out your Makey Makey earth connection. Fold a piece of paper in half, and cut out an aluminum rectangle that will just fit inside one half of the paper. To create earth, you'll need to lay down a piece of conductive tape and then tape aluminum foil on top of it. Remember from the paper circuits in Chapter 4 that the bottom of the conductive tape is not necessarily conductive because the adhesive may not be conductive. For this reason, make sure that the top of your conductive tape will touch the bottom of the aluminum foil, as in

Figure 8-16   Laying the groundwork.

Figure 8-16. Use double-sided tape to adhere the foil to the base of your paper circuit.

Step 2: Make Space for Switches

Grab another piece of paper, and cut it in half so that it will fit inside the folded paper you are using for your circuit. Use a pencil to draw out five or six spaces for your Makey Makey switches, as in Figure 8-17. Cut out each space. You can fold your paper in half to cut decent circles.

Figure 8-17   Draw a switch and cut it out.

Step 3: Get Switchy with It

Using your switch template, mark the top half of your paper. This is where you will put the switches that you will program with Scratch and Makey Makey. Cut small squares out of foil for each switch. Lay them out on your paper circuit background, and make sure that they will not touch one another because you want each switch to function as its own button. Do not tape them yet. Instead, place them to the side so that you can use conductive tape to give each switch its own personality (Figure 8-18). Remember, if they touch, then it will activate multiple keys on your Makey Makey at once—somewhat like pressing two keys on the piano at the same time.

Step 4: Routing Begins

Remember our branched paper circuit project in Chapter 4? This layout will be similar because we will make individual routes for each Makey Makey circuit. However, the way we trigger each switch will be different. Just remember that each switch needs its own route or you will have both activate at once (just like your LEDs would have activated at one time if your circuitry connected in Project 16). Lay out the first switch, which is the furthest from the edge of your paper (see Figure 8-18). We are going to bring all routes to this left side of the paper. We don't have much space, so this is why we are starting with the furthest lead and then adding each switch as we go. If you are worried about accidentally crossing lines, draw all your routes with pencil first.

A note on tools: remember your copper tape tricks from Chapter 4? A lot of the same tricks work well with this conductive soft tape from Makey Makey. However, this tape is different because you cannot tear it like copper tape, but this also means that this tape is more stable and solid for a project such as this one. The Makey Makey tape is more like a fabric tape, which

Figure 8-18 First switch route.

makes it a really cool material for marble walls, sewing circuit projects, and more!

Lay out all routing for each switch, and end all circuit routes on the left side of your paper. Remember that to make a corner with this tape, you'll bend back the tape to crease it as you did in Chapter 4. This will create a nice clean angle.

Continue this method until all switches have tape routing that leads to the left end of your paper circuit. Figure 8-19 shows how I laid out all of my routing.

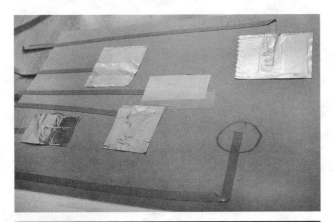

Figure 8-19 Routing for paper circuit.

Step 5: Add the Contact Plate

Now gather your aluminum foil squares because they are taking on a new life as a Makey Makey contact plate. Starting at the right-most switch spot, start adding these foil covers. Apply double-sided tape to the foil on the sides, but make sure that you leave a space in the center without tape for your conductive routing on your paper. Otherwise, you would accidentally insulate your switch because the adhesive on double-sided tape is not conductive. Therefore, be careful about your tape application.

Press and smooth your foil over the conductive tape as in Figure 8-20, and you've made your first Makey Makey paper circuit switch!

Continue making switches in this manner, being careful not to cross connections with other foil contact plates or conductive tape traces. You can easily place double-sided tape on your foil switch cover and place it on the paper, as in Figure 8-21, to make sure that it doesn't interfere with your conductive tape. Then attach the other piece of double-sided tape next to the conductive tape, and push your foil down to attach.

Step 6: Insulate and Attach!

Sometimes you might have your foil contact plates very close together. If this happens, you can insulate your switches by placing regular tape between them to insulate the connections.

Now it's time to attach our alligator clip test leads to make sure that your Makey Makey paper circuit works! But first remember to tape your insulator page between the two connections to keep them from being constantly connected. The orange paper acts as an insulator in Figure 8-22.

Attach an alligator clip to each routing for each aluminum foil switch, as in Figure 8-22. Attach the opposite end to an input on the Makey Makey. Connect to all the arrow inputs and spacebar, and you can even use the w, a, s, d, f, and g sockets (pins) on the back of the board by using the white jumper wires (also known as *connector wires*) that come in the kit. They are male-to-male wires. One end you will poke into the designated key (for our project we used the

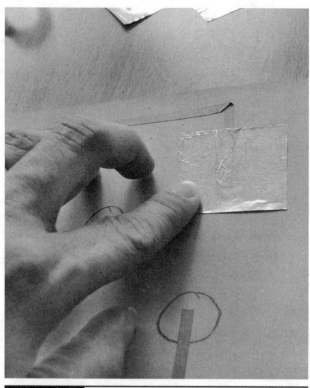

Figure 8-20   Make a switch.

Figure 8-21   Double-sided tape trick.

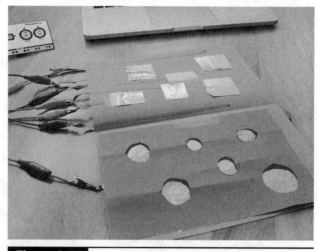

Figure 8-22   Insulate with regular tape and attach alligator test leads to your paper circuit.

w socket), and then you connect an alligator clip to the other end and connect the alligator clip to your paper circuit.

Don't forget to attach a lead to your earth switch! We found that our earth switch lead was too close to our other switches, so we eventually moved it to the bottom of our paper circuit, which you'll see in the next step.

Because you have a lot of leads, you may want to tape them down with regular tape for your test. Press down on each spot to see if the Makey Makey lights up. If they all work, it's time to trace your midi pads/buttons!

Step 7: Trace Buttons

Hold your paper circuit up to a light source, as in Figure 8-23, so that you can trace where all your switches are hiding when the card is closed. This will help you when you add flair in the last step.

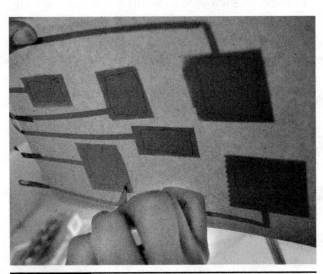

Figure 8-23 Trace.

Step 8: Move Earth If Needed

If your earth on the left side is connecting to one of your switches, you may need to move it. We ended up taking one of our white jumper wires and taping it under the foil on the earth side of our paper circuit and then clipped an alligator clip to earth, as seen in Figure 8-24.

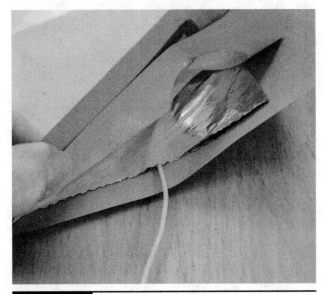

Figure 8-24 Move earth if needed.

Step 9: Label Buttons

Now that you have all your buttons and switches working properly, let's label them. Test each button on your paper circuit, and label which Makey Makey pad it is connected to (Figure 8-25). Once you've labeled all your switches on your paper, it's time to program some beats!

Figure 8-25   Connected and labeled.

Step 10: DJ Scratch

It's a good thing you went through those Scratch projects in Chapter 5! Now that you know how to program, you can program real world things with your Makey Makey. Even though we aren't really using the animation portion for

this project, feel free to add a background and make your paper circuit a controller for a game once you finish this project. As you learned in Chapter 5, you should always pull out a "When flag clicked" block to start every game. Attach the coordinates where you want your sprite to start, as you did in the coding projects from Chapter 5.

Classroom tip: This is a fun project for students who haven't used Scratch much before. It will teach them to program different arrow keys. Plus, who doesn't love tinkering with music and sound effects? So if you missed Chapter 5, feel free to introduce Scratch with this project; it is a very simple game for a first program!

Step 11: Space Oddity

Let's get scratching, DJ! Program your first switch by dragging out a "When space clicked" from the "Event" blocks and attach a "play sound" from the "Sound" blocks (Figure 8-26). For fun, let's record our own sound. Click "Record," which will show up when you click the downward-facing arrow on the "Sound" block.

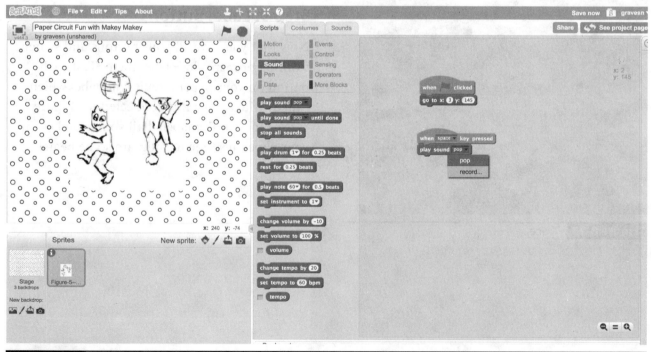

Figure 8-26   Program space key.

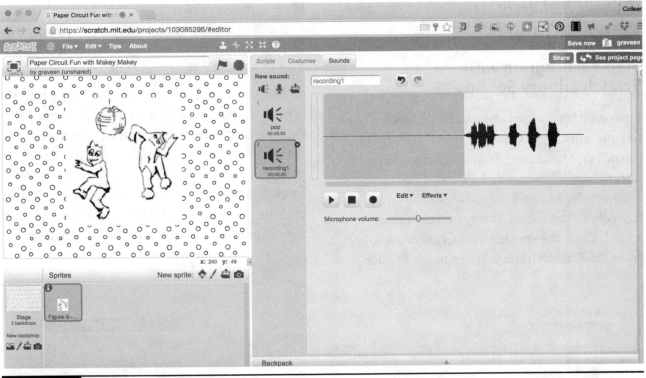

Figure 8-27 Record and edit.

Make sure that you allow "Adobe" access to your microphone so that you can record your own sounds.

Record your own sounds by pressing the circle, and press the square to stop the recording, as in Figure 8-27.

Scratch is pretty awesome because you can edit the sound and add effects all within this program! Just highlight the part you want to delete, and press Delete (see Figure 8-27). You can also highlight a section of your recording and add an effect by choosing one of the effects listed in Figure 8-28. I reversed my sound and then copied and pasted it.

Classroom tip: Students get really excited about creating their own sounds in Scratch. If they want to move around the room to get cleaner sound effects or use an external microphone, let them. If you have Synth littleBits, now is a

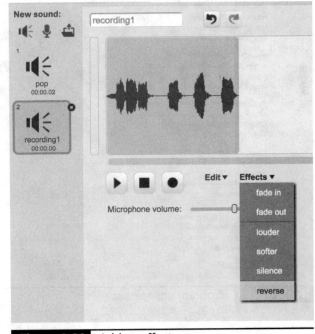

Figure 8-28 Add an effect.

great time to break them out so that students can upload their own electronic music right into Scratch!

Step 12: Program with Scratch Library Sounds

Drag another "When space clicked" block to your work area. Change "space" to "right arrow," and to choose from the Scratch sound library, you'll have to click "Record," as in Figure 8-29. Figure 8-30 shows you how to navigate to the sound library and choose a sound. Click the speaker icon to choose a sound from the Scratch library. When in the Scratch

library, press the play arrow next to each sound to listen to it. Once you find a song or sound you like, make sure that the song is highlighted as in Figure 8-30, and then press OK.

The sound will then show up in your editing area as if you recorded it yourself. You can edit the sound or leave it as is. Click on "Scripts" to get back to your script workspace. Make sure that your sound is loaded onto the right arrow by clicking the arrow inside the sound block and clicking on the sound, which is now located in the drop-down menu on the "Sound" button to program the right arrow.

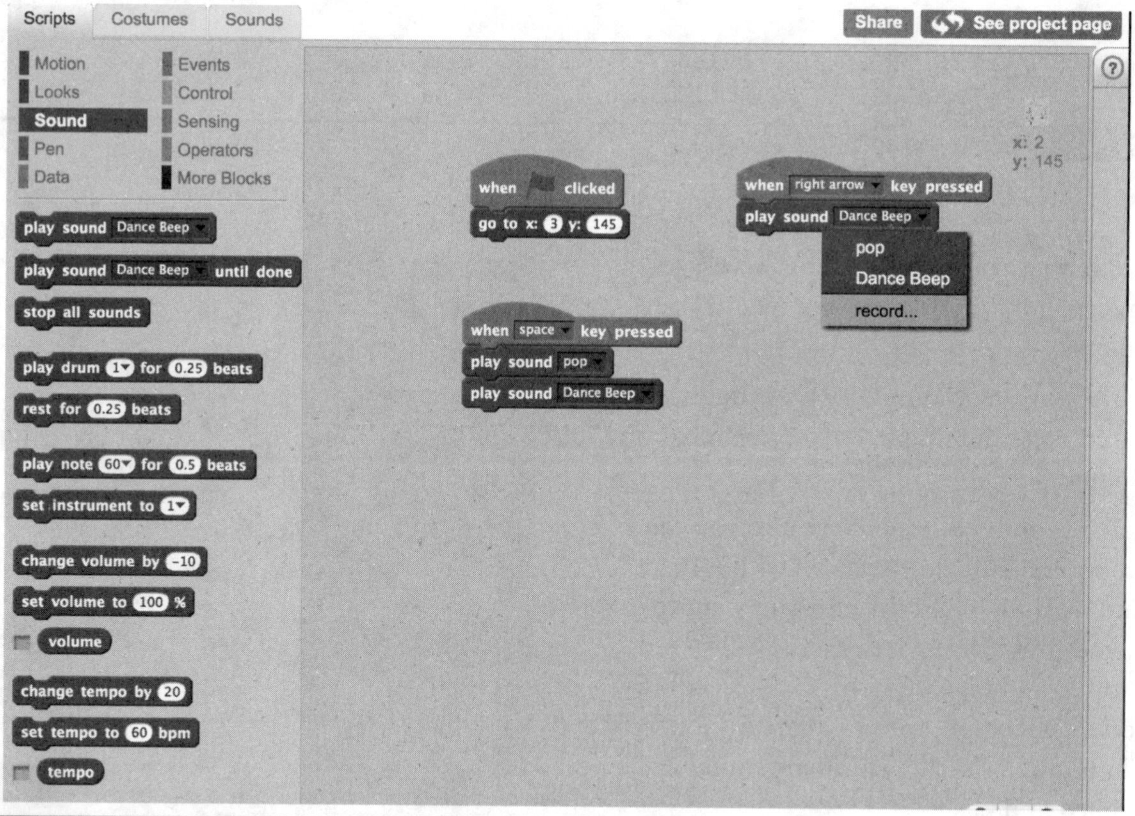

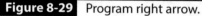

Figure 8-29   Program right arrow.

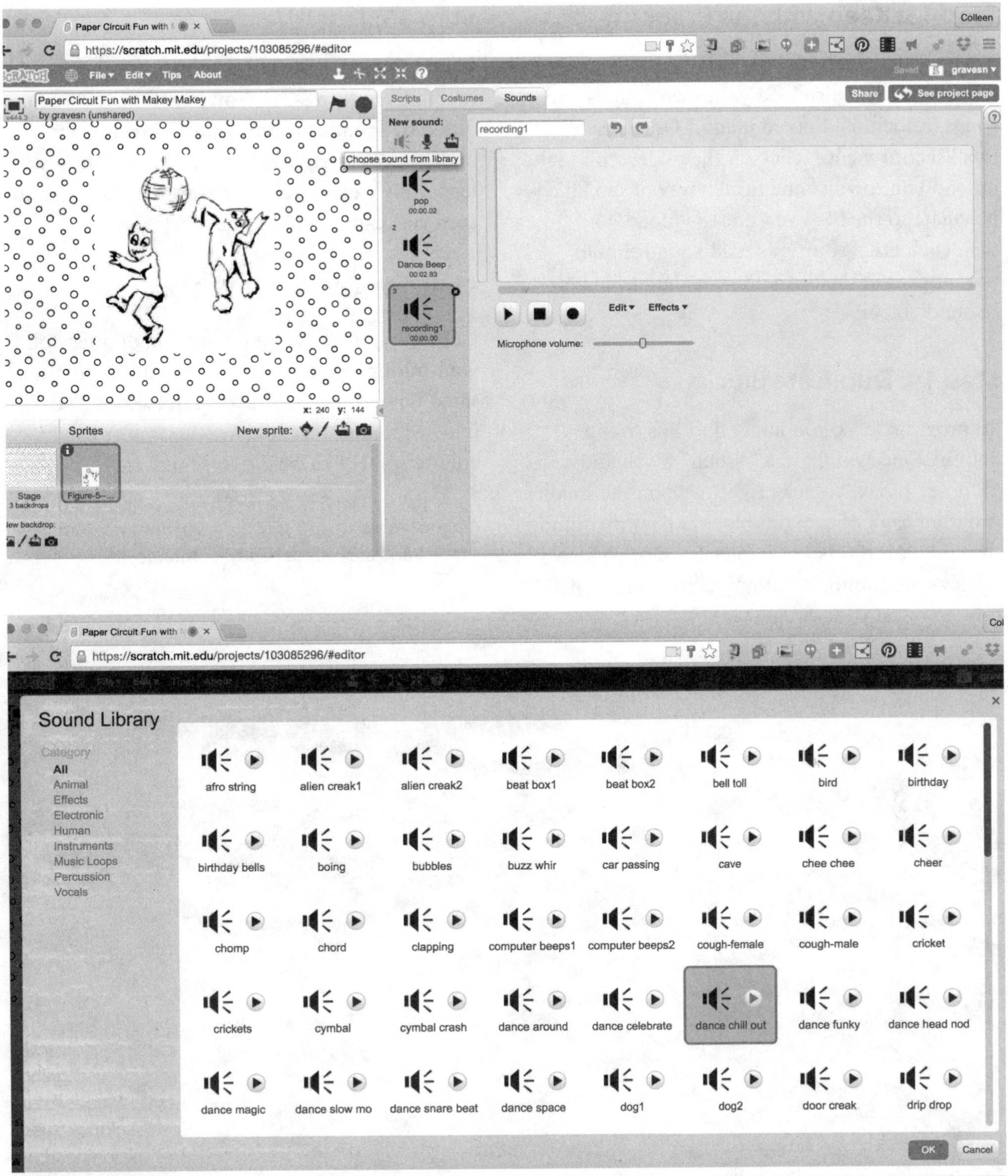

Figure 8-30 Choosing sound from Scratch library.

Step 13: Keep with the Program

Keep programming keys and tinkering with sound. You can also load your own sounds and songs inside the "Record menu." Once inside the "Record menu," click on the folder icon to upload your own sound from a wav or mp3 file. Navigate to the file you want to upload, and then click OK. Your uploaded sound should now appear in the drop-down menu inside the "Sound" block.

Step 14: Duplicate Blocks

To program keys, you might find this trick helpful. Once you have a "When" block and a "Sound" block, you can right-click on the small program to duplicate this block set as you did in the coding projects in Chapter 5. Once you have all keys programmed with different sounds, it's time to retest your circuitry!

Step 15: Test It!

Play each key to see how well your sounds work together (Figure 8-31). You may find you want different sounds once you start playing your midi paper circuit. Tinker with sounds until you have your paper circuit the way you want it, and then get ready to add some flair!

Step 16: Add Some WOO!

Cut out pictures or draw some fun doodles for your midi paper circuit. Attach with double-sided tape. If you want, you can even add some dimensionality by adding pompoms or bright color cutouts like we did in Figure 8-32.

Classroom tip: Keep lots of old magazines, random stickers, and old photos handy for this project.

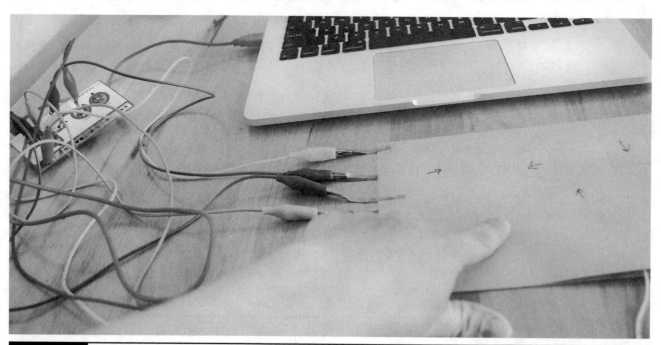

Figure 8-31   Test your circuitry!

Figure 8-32 Add flair.

Step 17: DJ Scratch Is in the House!

It's time for you to be the DJ! Click the green flag, turn up the volume, and get jiggy with it! Don't hog your paper midi; let others join the fun. Once you get tired of the song, reprogram your sounds, and play again! Rinse and repeat (Figure 8-33).

Challenges

■ Now that you know how to do it, can you make a giant paper piano?

■ How about a large-scale NES controller to play your favorite games with your feet?

Figure 8-33 Be the DJ.

■ What type of sound effects can you make with littleBits? What sound effects can you invent with things that surround you every day?

Project 35: Marble Wall Switches

Cost: Free–$

Make time: 10–15 minutes

Supplies:

Materials	Description	Source
Marble wall	Existing marble wall or cardboard tubes, pool noodle tracks, etc.	Build it
Invention kit	Makey Makey, alligator test leads, USB cable, and wires	JoyLabz
Conductive stuff	Foil, 5/8-inch steel ball bearings, copper tape, or soft conductive tape in Makey Makey Inventor's Booster Kit	Grocery store Amazon SparkFun JoyLabz
Nonconductive supplies	Cardstock paper, masking tape or clear tape, cardboard	Craft store Recycling bin
Tools	Scissors, box cutter	Craft store

Step 1: Roll-Over Switch

For this simple switch, start by cutting two pieces of foil 1 inch wide and about 4 inches long. Marble wall tracks vary greatly, but you can use this switch on most tracks as long as you are using a metal ball for your marble. It is essential for the strips of foil to be very close for this switch to work, but don't place them too close or your Makey Makey will be playing continuously. Place the two pieces of foil as close together as possible.

Let's test your "marble" before putting it on the track. Clip the alligator leads one to earth and one to the up arrow input on the Makey

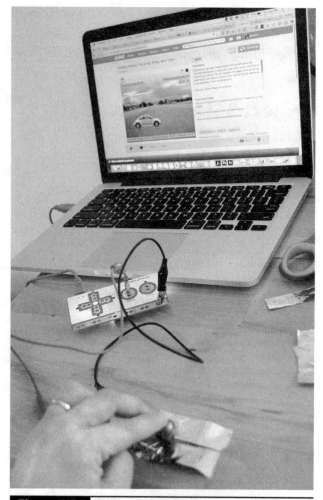

Figure 8-34 Foil strips taped and tested.

Figure 8-35 Testing the roll-over switch.

Makey. Then roll your metal marble over the track and make sure that it will activate the Makey Makey, as in Figure 8-34. If so, you are ready to place it on the track.

Use two small pieces of tape at the ends to hold the foil switches in place. Remember, if you cover the foil, the regular tape acts as an insulator, so don't cover all the aluminum foil, just use enough to make it stay on the marble wall. If your track has a square or round groove, fold the pieces of foil over the track, and position them with a slight gap in the center. Attach the foil to the track with tape, but leave some of the strip of foil hanging over the edge so that you can attach a wire or alligator clip test lead (Figure 8-35).

Connect a test lead to the earth input on the Makey Makey and then to one side of the switch. Clamp a lead to the opposite side of the switch and to the desired key input on the Makey Makey. Roll the metal ball bearing across the switch, and make any adjustments necessary to make it complete the circuit. You know it is completing the circuit when you see the light by the desired key light up, as in Figure 8-35.

Step 2: Repeating Switch

Sometimes you may want your marble wall to repeat the same sound or action, like a chime or an alarm ringing. In this case, you would want to use a repeating switch. This switch on the marble wall almost acts like a pulse because it will activate your Makey Makey in equally spaced intervals if you create your repeating switch in equal spaces. Start by cutting a thin strip of foil or length of copper tape to rest in the bottom of the track. Go ahead and place it inside the channel of your track. Be sure to leave a little extra to attach the earth wire to. Next, fold some strips of foil about 6 inches long and ½ inch wide.

Fold over each of these strips at about the halfway mark to make an L shape. Now attach the L's together with a piece of tape, as in Figure 8-36. Make sure that you wrap the tape all the way around so that the switch doesn't pull apart.

You will need something to suspend the switch over the path of the ball bearing. We used 1-inch pegs to suspend our switch, but there are lots of other materials you could use for supports: large paper clips, clothespins, pencil stubs, etc. Because this is a support and not part of the switch, it doesn't matter if your support is conductive or nonconductive. If you are using a paper clip, simply unfold it and tape it to the wall of your track to make a support for your switch. Once you have the switch in position, trim and fold the ends to fit on the track so that they will make contact with the ball. You might consider making a point at the end of each arm of the switch to help it return to its resting position. When the ball travels down the track, it will connect with the conductive tape you laid inside the channel and connect this with the foil L's suspended from the wall, thus completing the circuit.

Once you feel good about the position of your repeating switch, tape the switch to the supports. Clip one end of an alligator clip test lead to the strip of copper tape or foil that was placed in the bottom of the track, and attach the other end to the earth input on the Makey Makey. Run another alligator clip test lead to the beginning of the switch seen in Figure 8-37 and the other end to the desired key input on the Makey Makey (Figure 8-38). Use a ball to test the switch, and make any adjustments necessary so that the Makey Makey is triggered at every downward point.

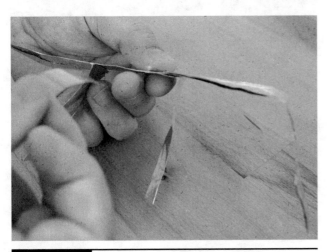

Figure 8-36 Taping L-shaped strips together.

Figure 8-37 Testing repeating switch.

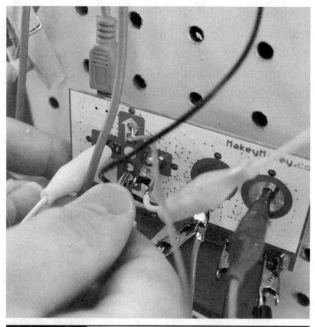

Figure 8-38 Plug alligator test leads into Makey Makey.

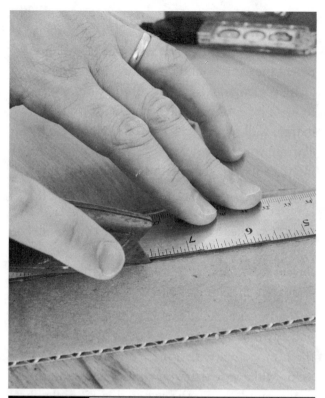

Figure 8-39 Scoring cardboard.

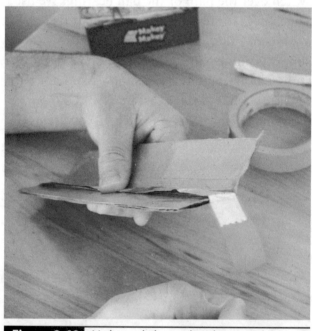

Figure 8-40 V-shaped channel with ground.

Step 3: Sequential Switch

This switch allows you to trigger several different actions in a sequence as the ball rolls down the track and over your sequential switch. The size of this switch and the number of items triggered can be changed to suit your needs. For this example, we will start by scoring cardboard as in Figure 8-39. After scoring, fold the middle of a piece of cardboard that is 3 × 10 inches (change dimensions to fit your marble wall). Fold the cardboard along the score to make a V-shaped channel. If you cut too deeply or just want to decorate the switch, use some duct tape to cover it.

Fold a strip of foil about ½ inch wide and about 12 inches long. Tape the foil strip about ⅛ inch off the center by folding the excess length over and fixing it to the back of the track (Figure 8-40). This strip will act as the ground, so you want to make sure that the ball will touch it as it rolls over the full length of your V-shaped channel.

Now let's make some switches! Cut and fold four rectangular foil shapes about 1½ × 2 inches. These strips will be the contact surfaces for each sound or action triggered by our DIY sequential switch. When you place these switches, you will need to consider how fast the ball will be rolling and how far apart you want the sounds or actions to occur. Our ball rolls relatively slowly down the channel because this is at the beginning of our track, so we went with a ½-inch gap between contacts. Use some double-stick tape on the bottom of your foil to place the strips about ⅛ inch from the center; then wrap the excess foil length around the cardboard and tape it in place, as seen in Figure 8-41.

To keep the edges of the foil strips from peeling up and to make the path so that the ball doesn't jump, we added some tape between the individual triggers. Now you are ready to attach an alligator clip test lead from the long strip of foil that runs the length of the track to the earth input on the Makey Makey. Now attach an alligator clip test lead to each contact switch and each input on your Makey Makey that you plan to program in Scratch. To create this sequential switch, you will really just create a series of individual switches that share a common earth. Each switch will have an individual alligator clip test lead or wire running to the corresponding key you want to trigger via the Makey Makey. When you test the switch, adjust the angle of the entire switch to allow the ball to roll slowly, allowing it to trigger each switch (Figure 8-42). Once it's tested and ready to go, place it on the marble wall, and test all your switches at once (Figure 8-43). Now you are ready to program!

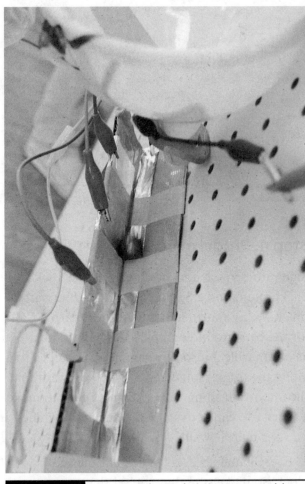

Figure 8-42 Sequential switch testing on marble wall.

Figure 8-41 Placement of contact surfaces.

Figure 8-43   Testing all marble wall switches.

Step 4: Program Scratch

Sign into Scratch and start a new game. Pick a background and a Sprite for your marble wall timer. We uploaded a picture of our marble wall for the background and picked a ball to function as our marble. I also programmed the ball to roll along the track. You can do this if you want, but the important thing here is learning how to make a timer you can activate with your new switches on your marble wall!

Every game needs a "When flag clicked" block, so make sure that you set that in your work area first and place your sprite where you want it to start at the beginning of your game. We programmed ours to start at the beginning of

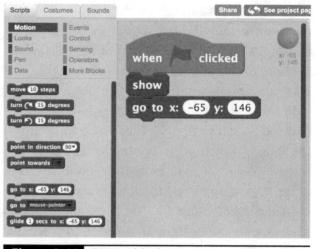

Figure 8-44   "Go to" block.

our marble wall. One of the easiest ways to get these coordinates is to drag your sprite to where you want it to start. If you have your "Motion" blocks open, you'll notice the x, y coordinates changing as you drag your sprite into the game area. You can use this to your advantage by placing your sprite and then dragging the "Go to" block to your script area, as in Figure 8-44.

Now you need to create variables inside the "Data" blocks so that you can control your timers. Make a variable named "milliseconds" and a variable named "seconds," and if you desire, you can add "minutes" (Figure 8-45).

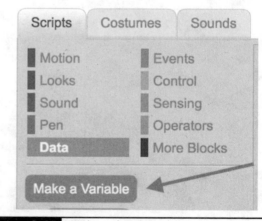

Figure 8-45   Make a data variable.

Once you've made your variables, you need to drag a "Show variable" block for each variable that you've created. Your variable timers will appear once you click the flag, as in Figure 8-46, but you may want to hide them later, so drag two "Show variable" blocks to your "When flag clicked" program. While you are here in the "Data" blocks, you also need to drag a "Set variable" block for each variable you created. You should now have three timers set to zero:

milliseconds, seconds, and minutes (Figures 8-47 and 8-48).

Now you'll want to go to the "Events" blocks and drag a "When space clicked" block to your work area. This block will program a key, so it does not connect to your "When flag clicked" block.

Using the drop-down arrow inside the block, change this block to read "When up arrow key

Figure 8-46   Click to activate timers.

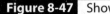   Show and set variables.

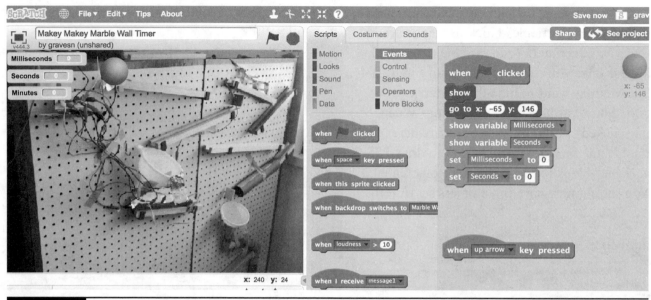

Figure 8-48   Timers showing.

pressed." This program is going to start the timer at the top of your marble wall. Remember the first Makey Makey switch we hooked up in Figure 8-35? When you place a metal marble here, it will now activate the switch and start the timers! However, we haven't told our timers how to function.

Drag a "forever" loop out of the "Control" blocks and connect it to the "When up arrow

key pressed" block. Open the orange "Data" blocks and drag a "Change milliseconds by 1" block inside the "forever" loop. As long as our timer is going, this will increase the time by 1 for every millisecond. However, what about once the timer reaches a second? How will the timer know to reset? We have to program it! So find the "if/then" statement inside the "Control" blocks, and nest it inside the "forever" loop, as in Figure 8-49. You'll find the "__ = __" operation inside

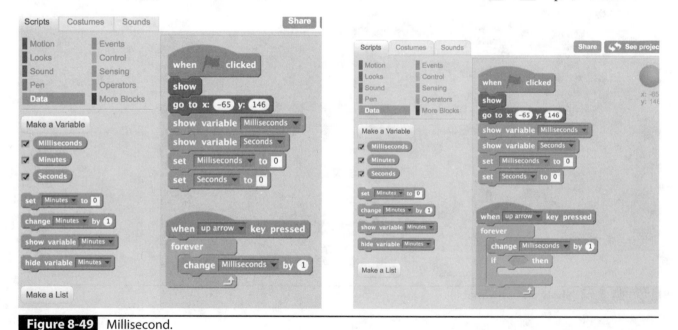

Figure 8-49   Millisecond.

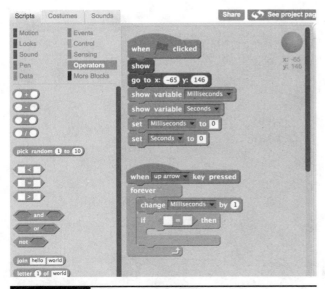

Figure 8-50 Call me operator.

the "Operator" blocks, as in Figure 8-50. Drag the "milliseconds" variable from the "Data" blocks into the first place in this equation. In the second place, put the number 1,000 because there are 1,000 milliseconds in 1 second! Almost done with our first timer! Now you'll want your milliseconds to reset every time you've reached 1 second. So drag the "Set milliseconds to 0" inside the "then" section of your "if/then"

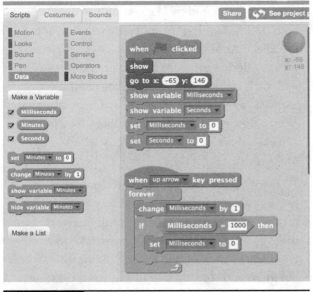

Figure 8-51 Milli Vanilli timer.

statement. You can see this program built in Figure 8-51.

Now it's time to program the timer to count seconds, and we are going to program it to activate with the up arrow key. From the "Events" blocks, drag another "When up arrow key pressed" block to your script area, and from the "Control" blocks, attach a "forever" loop. You'll want to change the seconds timer by 1 just as you did for the millisecond timer. Drag the "Change seconds by" variable from the "Data" blocks, and place it inside the "forever" loop. You'll have to find the "__ + __" operator inside of operations to place inside your "Change seconds by" block (Figure 8-52). Almost there! But what differentiates this from the milliseconds? Not much! We need to fix that before we move on to adding the minute timer. Drag a "wait 1 sec" block from the "Events" block, and attach it to the "change variable" block inside the forever loop.

As you did in the preceding timer, you'll want an "if/then" statement inside your "forever" loop. You need to add an operator to tell the computer what to do when "seconds = 60." Inside if the "then" portion of your "if/then" statement, set the seconds to 0 and change the minute timer by 0 + 1. This program is telling your computer to count 60 seconds and then mark 1 minute and start the seconds timer over (Figure 8-53).

Now you can program your other Makey Makey switches to make sounds, add more timers, or animate your screen. It's up to you! But wait! Before you go, you have one last program you need to create to make your timer really effective. A "Stop" button! Drag a "When down arrow pressed" block to your work area. We used a down arrow, but you can use whatever input you plugged your last Makey Makey switch into. Inside the "Control" blocks, you will find the "Stop all" block, which will stop

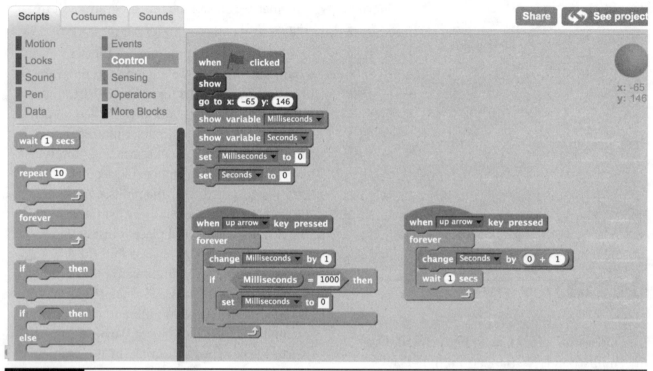

Figure 8-52 Hang on a second.

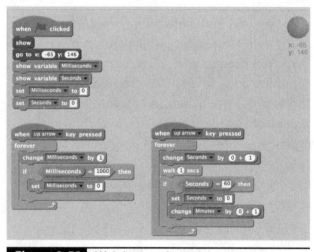

Figure 8-53 Wait just a minute.

your timers (Figure 8-54). Record your marble time, click the green flag to reset your timers, and play again! Figure 8-55 shows you our whole program. We decided to make it look like our marble sprite was gliding down the track as the real marble goes down the track! Adjust as necessary for your own track and programming needs.

"Makey Makey Assistive Technology" Challenge

Now that you know how to program in Scratch and build all kinds of awesome switches with Makey Makey, let's take your knowledge and use it for good! Inspired by Mark Barnett of Geekbus, Tom Heck ran a program for teenagers similar to the EPICS program at Purdue, which is a service-focused design program for engineers. He worked with high school students to create controllers with Makey Makey for students with special needs. Now it's your turn; using your new skills, can you create assistive technology with Makey Makey?

Once you've made your Scratch game and assistive technology with Makey Makey, take pictures of your challenge project. Tweet them to @gravescolleen or @gravesdotaaron, tag us on Instagram, and include our hashtag #bigmakerbook to share your awesome creations on our community page, which will host photos of you and your makers' projects.

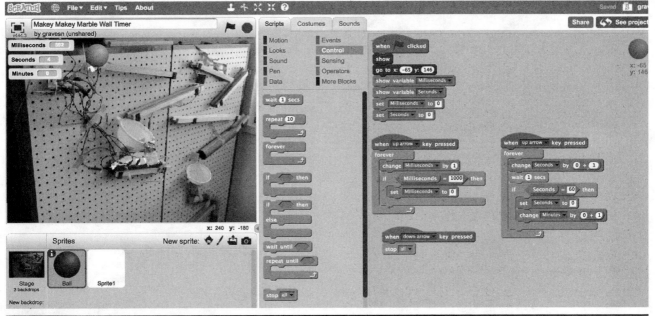

Figure 8-54 Stop! It's marble time!

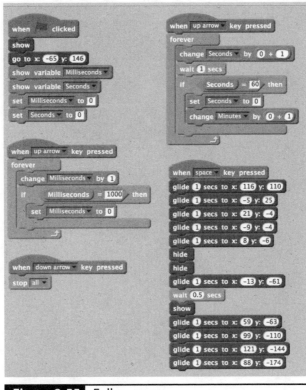

Figure 8-55 Full program.

Coding and Programming Objects

In this chapter we will start with some very basic programming with plug and play robots. As the chapter progresses, we will move on to more advanced coding for programmable objects and creating your own robot that you can code.

Project 36:	Programming Dash and Dot with Tickle App
Project 37:	Programming Triangles with Tickle and Sphero
Project 38:	Programming Triangles with Sphero and SPRK Lightning Lab App
Project 39:	Programming a Robot Dance Party with Trickle
Project 40:	Programming with Hummingbird Robotics and Snap!

Chapter 9 Challenge

Get ready for an app-smashing programming challenge!

Project 36: Programming Dash and Dot with Tickle App

The Tickle app is an amazing programming tool. It works with many different types of robots, so it's ideal for a beginner lesson into programming robots. Think of it like a Scratch for robots. Instead of a list of sprites, you'll have a list of robots as your characters! You can make games with it as well if you choose the "Orca," but we tend to use it daily with our robots.

Cost: $$$

Make time: 30 minutes

Supplies:

Materials	Description	Source
Programmable robot	Dash Robot	Wonder Workshop
Bluetooth-enabled device	iPad or iPhone	Apple
Programming app	Tickle app	Apple App Store
Tape	Low-adhesive masking tape	Hardware store
Tools	Tape measure, compass	School supplies

Equilateral Triangle with Dash/Dot

Step 1: Start New Project/Program

Open the Tickle app, and click the + sign to start a new project. When you start a new project, Tickle always asks you to choose a robot for your programming project. Look at all the different robots we can program! We can even choose to program a Philips Hue light bulb. Scroll down until you see Dash & Dot. Select it, and you'll notice it opens a prebuilt program. Each robot comes with an initial program, and they are all a little different, just like the robots themselves.

Step 2: Tickle Interface

Let's get to know the Tickle interface by running this first program and taking a look around. To see the initial program in action, tap the "Play" button above the line of scripts. The first thing you might notice is that the "Play" button turns to a "Stop" button. This is so because the program will be active until you tell it to "Stop." Even when the program is finished, you will still need to hit "Stop" or drag a new block to your workspace to stop running the program. You should also take note that as each line of code is performed, it lights up. Does Tickle remind you of Scratch? See how it has similar palette selections? Control blocks, motion blocks, sounds, looks, events, sensing, and even operators just like you did with Scratch. Browse through each palette of blocks to familiarize yourself with this app. Most of the blocks will be the same across robots, but some blocks will change as you choose different robots—especially the "Sensing," "Sounds," and "Event" blocks because those are unique to the functions of each robot. Your list of coding blocks is on the left, and your sprites (or robots) are at the top of your work area. You can get back to your list of projects by tapping the three lines at the top next to the program name. If you ever get stuck, there is a help menu located in the ? that takes you to an awesome list of how-to instructions. At the bottom of your scripts, you can even add variables, and at the bottom of your work area, you can see the "Undo" and "Redo" arrows.

Once you've run the initial program a few times, we are going to drag the bottom block to the left. A trash can appears in the script area. Pull the blue blocks out of the repeat block and throw them away. Just like Scratch

and Ardublock, you can pull blocks out of the code, or you can pull an entire script by dragging the first block of the program that you want to choose (whether it be for deleting or duplicating). We are going to scrap this whole program and start with fresh coding blocks, so drag the entire script to the trash (Figure 9-1).

We also want to add Dot to our work area, so click the "+Add" next to the robot list. Find Dot in the list. If there are multiple Dot robots in the room, you can choose your Dot by name by holding your finger on your robot. Once Dot appears in the sprite list, you'll see that it has its own program. Hit "Play" to see/hear what this program runs with Dot. Then drag everything except for the "When starting to play" block to the trash. Also, just as in Scratch, you will only see Dot's scripts when you are clicked on Dot, and Dash's scripts are hidden from view.

Figure 9-1   Delete.

When you click on Dash again, you'll see Dash's scripts, so don't worry; they are still there!

Step 3: Program Equilateral Triangle

For this first programming project, you are going to program an equilateral triangle because it will be easy to replicate the same angles and same distance for each side. If you want, you could set Dash on a piece of paper, attach a marker to it, see your robot draw out a triangle when you run this program.

If you threw away your "When" block, go to the "Events" palette and drag a "When starting to play" block to the work area. For fun, grab a "Change color of lights" block from "Looks." Because Dash is making a triangle, we can grab a "Repeat" block from "Control" and change the number to 4 by tapping the number to change it. To make Dash move, look for "Move forward" located in "Motion." With this block, you can adjust direction, time, and speed. You also need a "Turn direction" block to make the angles of your triangle. Where do you think a "Turn" block is located? You need Dash to move forward and turn three times at equal rates so that it completes an equilateral triangle. Normally, an equilateral triangle has angles inside the triangle that are 60 degrees. However, we are actually turning the robot from the outside angle of the triangle, so the outside wheel has to turn 120 degrees to turn your robot 60 degrees (Figure 9-2).

Step 4: Program Dot

Let's program Dot to play a sound at the start of the game so that it seems like it is cheering for Dash. Let's also program it to change colors by dragging a "forever" loop out of "Control" and putting the "Change color" block inside it. If you add a "Wait 1 sec" block to this script, it will give your Dot a flashing effect (Figure 9-3).

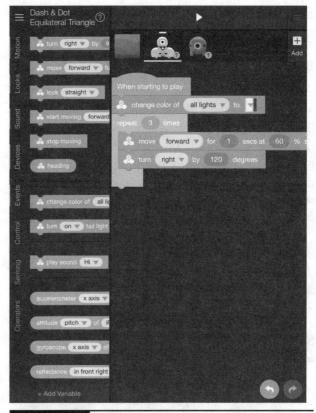

Figure 9-2   Dash program.

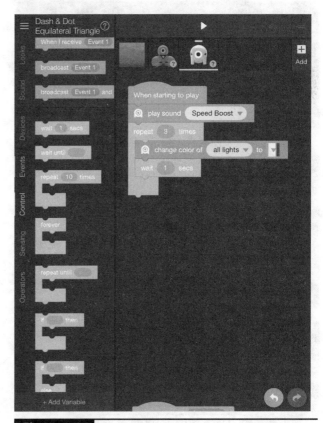

Figure 9-3   Dot program.

Step 5: Broadcast and Receive Messages

Because the triangle was an easy program to write, let's add a little broadcast message from Dash. Go to "Events" and find the "Broadcast" block and drag it to the end of Dash's scripts (Figure 9-4). Change the broadcast message to "I made it!" by clicking inside the script to type your own message. Now you'll need to drag a "When I receive" block to Dot's script area for Dot to receive the message. Look at Figure 9-5. If you were to run the program right now, what would happen? What is wrong with this code?

Step 6: Debug

Since we changed Dash's message to "I made it!," we have to choose the right message for Dot's "Receive" block. Figure 9-6 shows the correct scripts. Can you explain what was

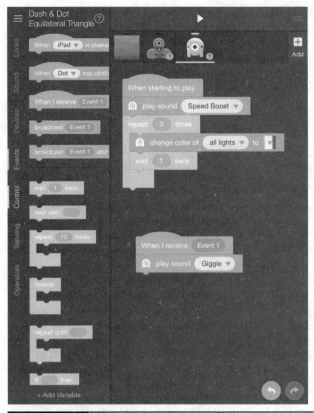

Figure 9-5   Dot receive: what's wrong?

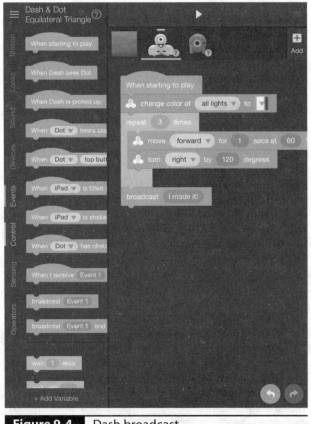

Figure 9-4   Dash broadcast.

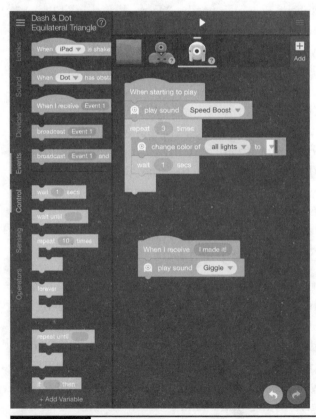

Figure 9-6   Debugged.

wrong with the code from Figure 9-5? If so, you are well on your way to adding "debugging superpowers" to your tool belt!

Right Triangle with Dash and Dot

The equilateral triangle is a pretty easy project. So let's try something a little more difficult and see if we can get Dash to drive a right triangle. Because it won't be as easy for us to see the program, you should lay out a right triangle with tape on a gridded floor. A tile floor should provide an easy programming grid!

Step 7: Tape the Right Triangle

First, we need to give Dash a starting point, as in Figure 9-7. If you don't give Dash a starting point, you won't be able to accurately program it do drive your triangle. Use a protractor to

Figure 9-7 Dash's starting point.

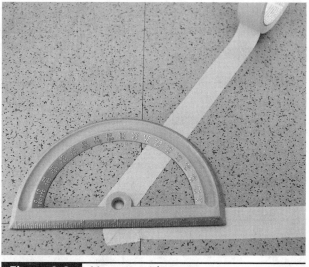
Figure 9-8 Measure with a protractor.

draw a triangle with tape on your floor. Because you are making a right triangle, your other two angles need to be 45-degree angles. Take note that the outside angle to a 45-degree angle is 135 degrees, as you can see in our measurement in Figure 9-8. You'll need this number to program Dash to turn at a 45-degree angle because remember that it is Dash's outside wheel that makes it turn. If you want students to do mental math, an easy way to find the number you need is to subtract the angle you want Dash to turn from 180.

Step 8: Write the Program

To start, we'll program Dash to move forward and turn left at 90 degrees. To program the next two turns, we may have to do a little adjusting. Drag a "Turn left by degrees" block from "Motion," and change the angle to 135 degrees. Drag the "Move forward block," and snap it to the "Turning" block. Since we want Dash to drive the same amount and turn at the same angle twice, we can place this inside a "Repeat" block, which you will find in the "Control" area. Can you see the error in Figure 9-9? What do we need to change to make Dash drive the triangle correctly? The angles from our program

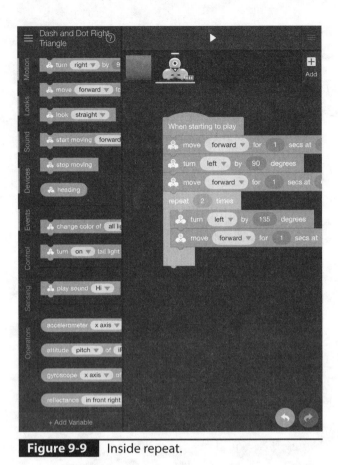

Figure 9-9   Inside repeat.

should work for your triangle, but you will have to adjust timing and speed for your own course. Different flooring types will affect friction and timing, so make sure that you factor those in! This exact same program with the exact same tape triangle on a carpeted floor will produce very different results! This project is more about you figuring out how to program your robot, so remember to adjust your variables as needed.

Step 9: Debug

You can't use the "Repeat" block for this program, so you'll need to model your scripts to look similar to Figure 9-10. Dash will start at one of the 45-degree angles, drive forward, and turn left at a right angle; then the outside wheel will turn 135 degrees to send Dash back to the starting point.

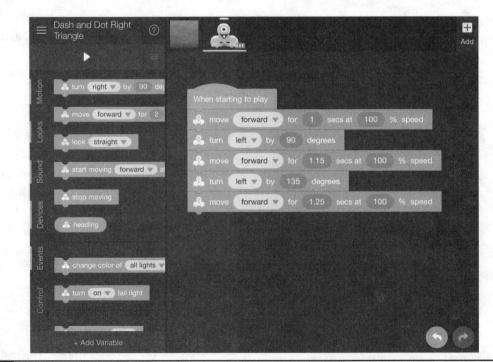

Figure 9-10   Debugged.

Step 10: Broadcast Reward

Let's reward Dash for seeing Dot! Put Dot at the starting point so that when Dash gets back to the beginning and sees Dot, he makes an airplane noise. Drag a "When Dash sees Dot" block to your script area, and add a "Turn on taillight" block from "Looks" and a "Play sound" block from "Sounds." Your program should look like Figure 9-11. Test the program. What happens? Why do you think this happens?

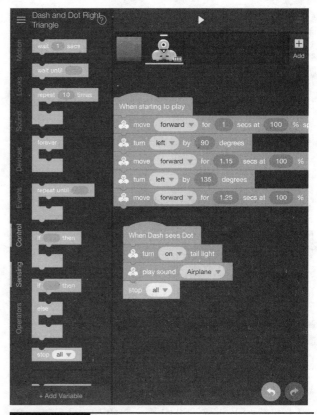

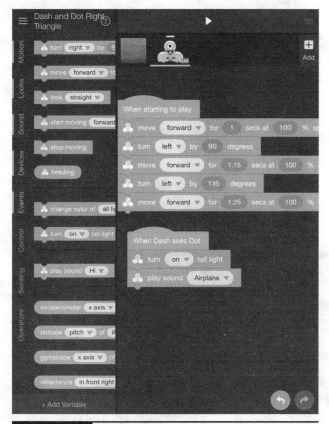

Figure 9-11 Reward your robot.

Step 11: Debug

Unless you program Dash to drive in another direction, it will continually make airplane noises as it looks at Dot. To stop this, we'll use another trick we learned in Scratch. From the "Control" blocks, grab a "Stop all" and attach it to this script, as in Figure 9-12. What other ways could you debug this program? Make sure that Dash

Figure 9-12 Debugged.

starts his drive facing away from Dot; otherwise, Dash will broadcast his message too soon. When it drives toward Dot, you'll hear the rewarding airplane noise that means that your program was successful (Figure 9-13).

Figure 9-13 Hi Dot!

Obtuse Triangle with Dash

Step 12: Obtuse Triangle Full Program

Now that you are getting pretty good at creating programs for Dash, see if you can create an obtuse triangle and adjust your variables so that it drives the obtuse triangle? Your program/scripts should resemble Figure 9-14.

Step 13: Test and Adjust

Because an obtuse triangle has one very long side, you may have to tinker with adjusting speeds and travel time. Change settings until you get Dash to drive your complete triangle (Figure 9-15).

Challenges

- Now that you know how to broadcast and receive messages, can you program a story or a conversation with Dash and Dot?

Figure 9-15 Start.

- What other ways could you program an interaction with these two robots?

- Check out the other apps provided by Wonder Workshop to learn more about this cute little duo!

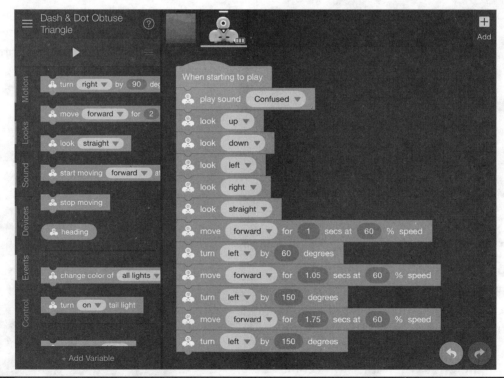

Figure 9-14 Obtuse triangle full program.

Project 37: Programming Triangles with Sphero and Tickle

Sphero is another prebuilt robot that is fun to program. It is a little harder to program shapes and obstacles with Sphero because of its round shape, but this just means that you'll build your perseverance superpower while you learn to program with Sphero. Just what is Sphero? It is a supercool robotic ball powered by a gyroscope and accelerometer. It travels at a much faster speed than Dash and does not have external wheels. Therefore, it makes a great experimental tool for physics! In the next two projects, you'll experiment with programming the same triangles from the preceding project using the Tickle app and learn to use the native Sphero app, SPRK Lightning Lab.

Cost: $$$

Make time: 30 minutes

Supplies:

Materials	Description	Source
Programmable robot	Sphero or Ollie robot	Orbotix
Bluetooth-enabled device	iPhone or iPad	Apple
Programming app	Tickle app	Apple App Store

Right Triangle with Sphero

Step 1: Wait for It

Try writing the same program for your right triangle as you did for your Dash robot. What happens? What does Sphero do at each turn? What can you do to fix this? Because Sphero is so much faster than Dash, the first thing you'll want to do is adjust your speed. We recommend going under 40 percent because we want to see Sphero correctly drive our triangle

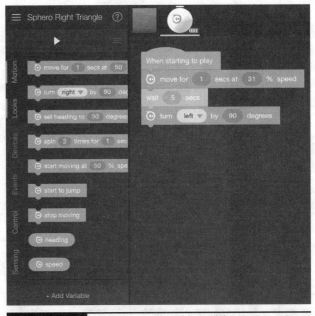

Figure 9-16 Adjusting speed by adding a "wait."

course. (Speaking of speed, what do you think would happen if we lower Sphero's speed to under 10 percent?) One of the fun things about programming is that there are many right answers to a single problem, so try other options you think of and see what happens. You'll also need to grab a "Wait secs" block from "Control" (Figure 9-16). Sphero needs this "wait" command so that it can stabilize before turning and moving.

Step 2: Turn

For the most accuracy, you should program your turn after the "wait" block, as in Figure 9-17. Sphero will still turn 90 degrees and head toward the next angle. As with the last project, the speeds and distances in your program will vary according to your driving surface. If you program your turn as in Figure 9-18, you'll see your Sphero error again because you did not add a "wait" for it to stabilize. This happens a lot in programming, so don't get frustrated; instead, use this as motivation to persevere! You'll become a better programmer for it!

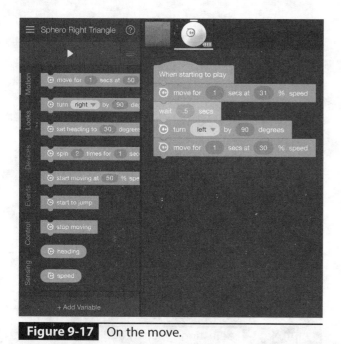

Figure 9-17   On the move.

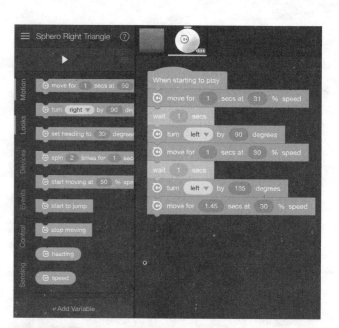

Figure 9-19   Full right triangle program.

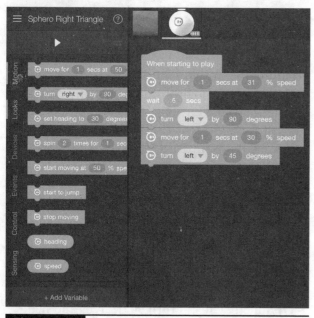

Figure 9-18   Turning. error

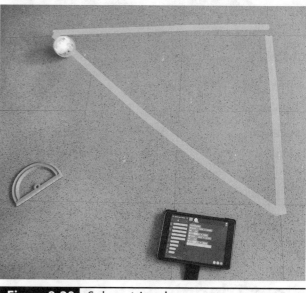

Figure 9-20   Sphero triangle.

Step 3: Full Program and Adjusting

Because Sphero is a round ball, it can also be hard to align it at the start of your program. Make sure that you align the tail light every time you start the program over. Figure 9-19 shows our full right triangle program, but again, yours will vary according to your driving surface and length of triangle. Figure 9-20 shows the Sphero program driving in real life!

Challenges

- Can you program Dash and Sphero to interact?

- Tinker with the light settings for Sphero. What can you learn about RGB coloring and percentiles?

- Could you design a game with this app for Sphero to play?

Project 38: Programming Triangles with Sphero and SPRK Lightning Lab

Tickle is supercool, but sometimes it's easier to program Sphero to drive a shape or an obstacle course with Sphero's native app, SPRK Lightning Lab. One of the coolest things about this app is that you have the ability to drive Sphero back to you after you test each program within the app!

Cost: $$$

Make time: 30 minutes

Supplies:

Materials	Description	Source
Programmable robot	Sphero or Ollie robot	Orbotix
Bluetooth-enabled device	Android or Apple phone or tablet	—
Programming app	SPRK Lightning Lab App	Apple App Store Google Play Store

Figure 9-21   Sphero start spot.

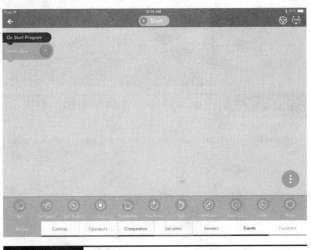

Figure 9-22   SPRK app.

Step 1: SPRK App

Download the SPRK app, and go through the tutorial. As we said earlier in step 3 of the previous project, Sphero can be hard to align. Create a start spot similar to Figure 9-21. Plus, if you have your start on a gridded floor, align the tail of Sphero with a line on the floor.

The SPRK app is similar to Tickle and Scratch but with some different command names for your palettes. Instead of "Motions" and "Looks," most of these controls are located within "Actions" in the SPRK app, as in Figure 9-22. However, you still have similar palettes such as "Operators," "Variables," and "Events." Across all drag and drop programming languages, these terms should stay the same.

Step 2: Set Heading

In SPRK, we will control our angles by adjusting the heading. Instead of being based on 180 degrees, Sphero will be programmed based on 360 degrees (Figure 9-23). You'll still want to use your protractor to measure your angle, as in Figure 9-24. You can drag the arrow to the heading you want Sphero to go. To begin, we will set Sphero at the start and program it to

Figure 9-23 Heading in app with Sphero.

Figure 9-24 Measure angle.

drive a 0-degree heading. This heading will drive Sphero straight forward (straight in relation to its tail light. Did you re-aim the tail light? You have to remember to do it at the start of every program run). If you wanted Sphero to reverse, you would need to set its heading at 180 degrees.

To turn right, you'd use a heading of 90 degrees, and to turn left, you'd set your heading to 270 degrees. At our next turn, you can measure your angle with a protractor and see that the angle of your triangle is 35 degrees, but to change Sphero's heading to drive in that direction, you'll actually need to measure this as if it were on the 180-degree scale. Our heading for Sphero at this turn is quite easy to determine because we can read that the opposite angle of 35 degrees is 145 degrees. Realize that you are not turning Sphero at an angle; instead, you are telling it to move toward the heading of 145 degrees to make the small turn toward your obtuse angle. When you get to your obtuse angle, the math is not quite the same nor as easy. Sphero operates on 360 degrees, so this means that we need a second protractor to measure the degree turn. To get Sphero to turn past 180 degrees and head back toward the original spot, we need to go 35 degrees past 180 degrees in order for it to arrive back at the start. This means that we need to program Sphero's heading to 215 degrees. The ideal tool for programming Sphero through an obstacle course or to make a shape is a 360-degree protractor. If you do not have one, you can use the app to try to determine your angle. Otherwise, guessing and experimenting are the best ways to adjust timing and angles to program your triangle (Figure 9-25).

Step 3: Wait or Strobe?

Instead of adding a "wait" block, why not program Sphero to blink a strobe light at each spot it needs to turn. (See what we mean about multiple ways to solve one problem?) The "strobe" block is located inside the "Actions" palette. You can choose a color on the color wheel by tapping the color you'd like your Sphero to blink. You can also adjust how many times you want it to blink and how long you want it to blink. In Figure 9-26, we've

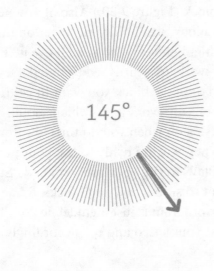

Heading

145°

Figure 9-25 Experimenting with angles.

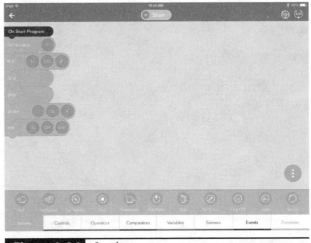

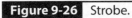

Figure 9-26 Strobe.

Figure 9-27 Strobe IRL.

programmed Sphero to blink the color pink four times in 1 second (see also Figure 9-27).

Step 4: Program

Because Sphero has three stopping points, let's program a strobe light at each stop, as in Figure 9-28. To set the strobe, you will choose the duration of the blink on and how many times you want the light to blink.

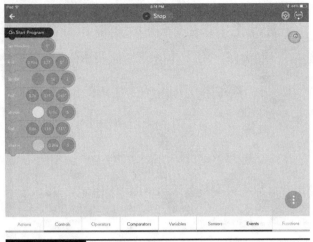

Figure 9-28 Full program.

When you set the "duration" (the middle number in the "strobe" block), this is like how you added a "Delay microseconds" in your Arduino program in Chapter 5. The duration you set is actually setting how long the blink will flash on. So, if you put in the duration to 1 second, your light will flash on for 1 second and off for 1 second. The second number in the block is how many times it will blink. We are going to set a different strobe at each turn to make Sphero wait, but we are also going to experiment with different duration lengths and different blink amounts. See Figure 9-28

Figure 9-29 Strobe at turns.

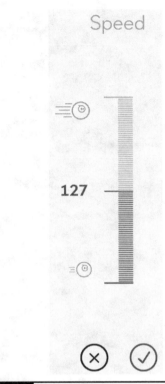

Figure 9-30 Setting speed.

for code and Figure 9-29 for Strobe in real life. Remember, you can set your speed in the "roll" command by changing the variable in the middle of the block (Figure 9-30.) Also, if you need to delete a block, just hold your finger on the block until it highlights, and you can move it. The top of your work area turns red to signal a trash can. You can drag the block you want to get rid of to the top of your work area (which should bring up a trash icon when you do this) to throw it away. When you are ready to test your program, press "Play" to run your program and "Stop" if you want to stop it and add blocks. Remember that our program is just a guideline. You'll have to adjust your speed and roll accordingly (Figure 9-31).

Challenges

- Can you program Sphero to drive an acute triangle?

- Can you build or design a chariot for Sphero? How would this affect your program?

Figure 9-31 Running program.

■ Try using different covers on your Sphero to test how the different levels of friction affect Sphero's speed.

Project 39: Programming a Robot Dance Party With Tickle

Using the Tickle app, you can program your robots to dance together or even have a synchronized dance party. You'll notice that the Tickle interface is very similar to Scratch, so it's an excellent step to try after you have made a few successful games. Each robot will function like a Sprite in a Scratch game. You will need to write the following scripts for each robot you want in your dance party, but as with the other projects in this chapter, these scripts are merely guidelines. You will have to adjust your speed and movement accordingly.

For this project, we want our robots to dance together, spin, and then burst into spontaneous movements. You'll need to open the Tickle app on an Apple device and start a new project. This one project will control all the robots at once. Once you run the program, if your robots seem off, you will have to tweak your math depending on the surface of the robot's dance floor (Figure 9-32).

Cost: $$$

Make time: 30 minutes

Supplies:

Materials	Description	Source
Programmable robot	Sphero, Ollie robot, Dash robot	Orbotix Wonder Workshop
Bluetooth-enabled device	iPhone or iPad	Apple
Programming app	Tickle app	Apple App Store

Figure 9-32 Dance party robots.

Step 1: Program Dash Two-Step

Because Dash is the easiest, let's start with programming it first. You'll program Dash's color and have it move forward and back two times. To do this, scroll to "Events" to drag out a "When starting to play" block. Program Dash's color by getting a "Change color of all lights" from "Looks" and connecting your blocks. Remember, if you don't connect your blocks, the program will not run through every script. Now drag out a "Repeat __ times" block under "Control." Pull out a "Move for __ secs at __ % speed" and "Turn right by __ degrees" under "Motion," and stick them inside the "repeat" block. Change the number in the "repeat" block to 2 (see the full program in Figure 9-33).

Step 2: Program Dash to Spin

Programming Dash to spin is easy. You'll use a "Turn right by __ degrees" block and input 360 for the number of degrees. Program Dash to spin three times by cloning this block or by using a "repeat" block. We want Dash to stop on the last spin at 90 degrees so that we can program our last dance move. (Dash will look around while Sphero runs an arc around it.)

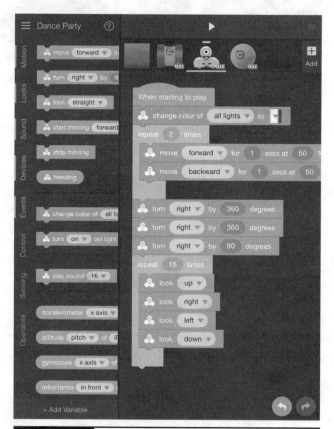

Figure 9-33 Full program for Dash.

Step 3: Program Dash's Head to Bobble

Under "Control," you will drag a "Repeat __ times" block and attach it to your spinning script. Drag four "Look straight" blocks from the "Motion" section and place them inside the "repeat" block. Change your "Look straight" blocks to program Dash to look up, right, left, and down. Change the number inside the "repeat" block to 15. Dash is now ready to dance! Figure 9-33 lists the full script for "Dancing Dash."

Step 4: Program Ollie to Two-Step

Next, we'll program Ollie. Click "+Add" to add a new robot for your project. We want to program Ollie to move back and forth the same distance as Dash and follow the same timing. Because it has a different motor, you may have to tweak some of the timing or the rate of speed at which Ollie performs each task depending on your driving surface.

Tap on the plus sign to add Ollie to your program and begin writing a script for it. Under "Events," find a "When starting to play" block to start Ollie's script. Grab a "Change color" block from "Looks," and attach it to the "When starting to play" block.

You'll find a "Repeat 10 times" block under "Control." Get a "Move for 1 sec at 50% speed" block, and place inside the "repeat" block. Change your seconds to 0.5 and the speed at which Ollie will travel to 40 percent. (See full Ollie program in Figure 9-34.)

Because Ollie's motor will continue to run if you don't tell it to stop, you need a "Wait 1 sec" block before having Ollie turn around and move back to its original starting point. Again, these are just guidelines for the movements; adjust the variables to suit your dancing needs.

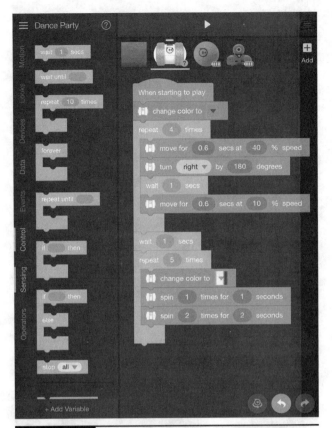

Figure 9-34 Full Ollie program.

Step 5: Ollie Backspins

Getting Ollie to do backspins is also pretty easy because there is a block for spins under "Motions." You'll want to grab a "repeat" block from the "Control" commands and snap it under your last "Wait 1 sec" block. To add a little flavor to your dance party, grab a "Change color" block from "Looks" and place it inside the "repeat" block. Now place two "spin __ times for __ seconds" blocks. You'll need to adjust all the variables inside these blocks to make it seem as if Ollie is dancing with Dash. The best way to do this is to run your program and see how they dance! It's best to do this at each step. If Ollie seems to be lagging, stop the program and change it (Figure 9-35). If you create the whole program and then try to fix small errors, it will take longer than fixing the dance errors as you go.

Figure 9-35   Testing robots spinning.

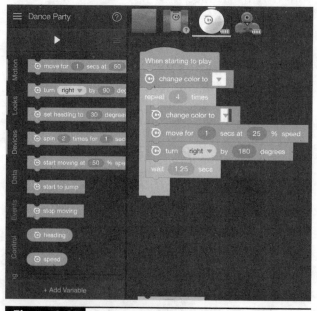

Figure 9-36   Sphero: test it!

Step 6: Sphero Two-Step

Click the "+Add" icon to add Sphero to this dance party. Try running the program from Figure 9-36. What happens? Remember from our triangle programs that Sphero needs a "wait" command to stabilize. The correct program to have Sphero dance the two-step with Dash and Ollie is in Figure 9-37, but remember that you will have to adjust the seconds and the speed percentage accordingly.

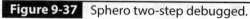

Figure 9-37   Sphero two-step debugged.

Step 7: Sphero Gets Dizzy

Programming Sphero to spin is also pretty easy because you can grab a "Spin __ times for __ seconds" block and place it inside a "repeat" loop. As with our "strobe" script in the last project, you'll set the duration of the command. However, this time, you are telling it how many times to spin and for how long. In our script, we said, "Spin 10 times for 1 second," so Sphero spins 10 times in just 1 second (Figure 9-38). This could be a very fun line of code to experiment with. Try spinning four times for 1 second. What happens? Now spin one time for 1 second. What do you think would happen if you told Sphero to spin 100 times in 1 second?

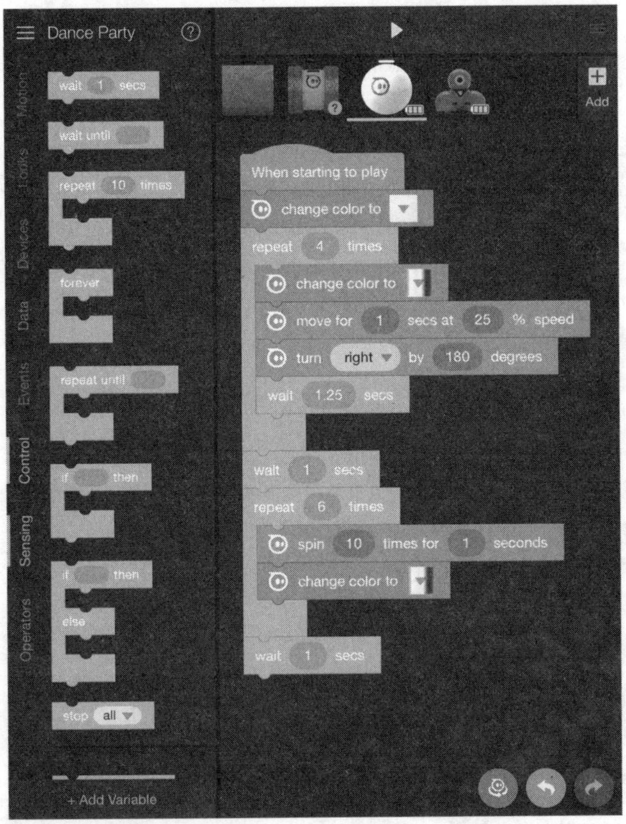

Step 8: Drive an Arc

Getting Sphero to drive in an arc is a little more complicated. You'll use an equation and your budding algebra skills for this fancy footwork.

Go to the "Data" palette to "Set X to 0." In your program, you are going to set an equation that repeats until the value of X is more than 180. To do this, you will program Sphero to move for 0.05 second at 40 percent of its speed. It is starting at a heading of 0 degrees, but as it rolls the value of X (also known as the *heading* in this scenario), it changes in 10-degree increments. When the heading goes over 180 degrees, it will stop.

To return, you'll repeat the program, but this time you repeat X until it is less than –180 degree. Figure 9-39 shows the full arc program.

Figure 9-38 Sphero spin.

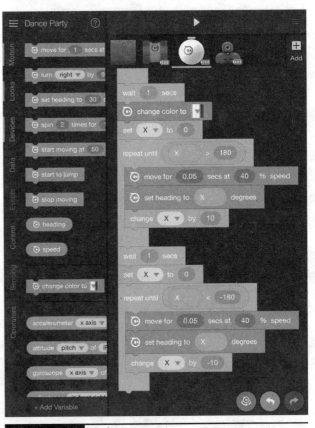

Figure 9-39 Sphero arc.

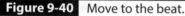

Figure 9-40 Move to the beat.

Again, this is a synchronous dance, and you may have to adjust the times. But start with this basic program, and then tweak it from there. Let's dance (Figure 9-40)! If you are having trouble with synchronicity, you may want to start your dance partners on a line on a gridded floor, as in Figure 9-41.

Challenges

■ Create a dance routine to your favorite song!

■ Create a program that makes one robot the lead dancer and the other robots backup dancers.

■ Create a dance-off between two robots, in which they try to mimic one another and then add a new step.

Project 40: Programming with Hummingbird Robotics and Snap!

The Hummingbird Robotics Kit is a fantastic way to introduce makers to robotics and Arduino microcontrollers. The microcontroller is Arduino at Heart, which means you can use your new Arduino programming knowledge. Plus, the kit includes all the parts you might want to build robots, an interactive sculpture, and more. You can easily combine LEDs, servos, geared motors, and recycling to create robots and interactive objects. Even better, there are

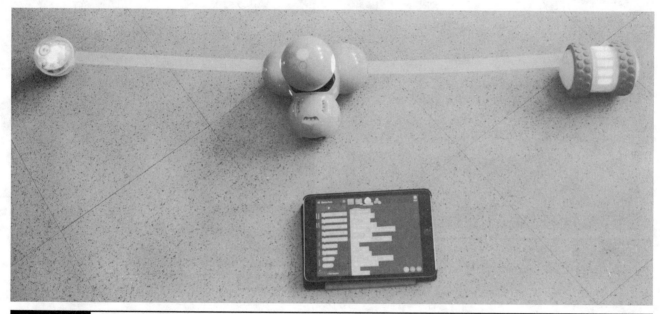

Figure 9-41 May need a start line.

several options for programming your new robot based on skill level and age. Students can use Snap!, a Scratch-like programming environment customized to work with Hummingbird robotics, Ardublock, and Arduino. The Hummingbird Duo board features easy-to-read ports and plug-in components, so even the littlest of makers can enjoy making robots. In this project you'll use the Duo kit to make a basic two-wheeled bot with LED indicators and program your bot with Snap! (Figure 9-42).

Cost: $$–$$$

Make time: 25–35 minutes

Supplies:

Materials	Description	Source
Robotics kit	Hummingbird Duo Premium Kit, battery pack (optional), USB extension cord (optional)	Bird Brain Robotics www.hummingbirdkit.com/
Computer	Computer with Internet access that meets the requirements to run Snap!	
Software	Snap! and BirdBrainRobotServer	www.hummingbirdkit.com/
Robot building supplies	Large craft sticks	Dollar store
	Tape	Craft store
	Container	Recycling bin

Step 1: Container and Motor Setup

Select a box or plastic container with square corners. Most plastic containers have a slight angle or curve at the bottom but still can be used. Locate the wheel adapters, rubber O-rings, and geared motors. Place a rubber O-ring on the wheel adapter, and then slide the wheel adapter onto the motor.

Figure 9-42 Two-wheeled Hummingbird bot.

Step 2: Attach Wheels and Motor to Container

Position the motor on the bottom of the container or box. If your container has a slight angle or curve, ensure that the motor is positioned so that the wheel has room to rotate, as in Figure 9-43. Use a large piece of masking tape across the gear box to hold the motor in place. Follow the contours of the motor when applying tape. Remove the wheel, and place a piece of tape at the front and back of the motor where it meets the container, as in Figure 9-44. You may need to use more than one piece of tape to secure the motor. Repeat the step for the opposite side once you feel that the motor is secure (Figure 9-45).

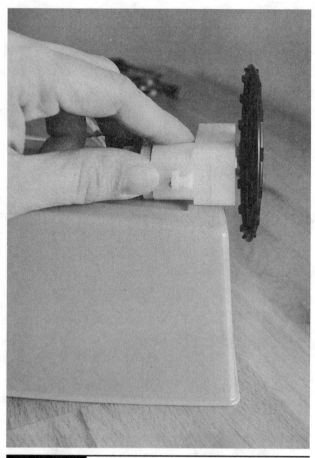

Figure 9-43 Checking for clearance.

Figure 9-44 Applying tape.

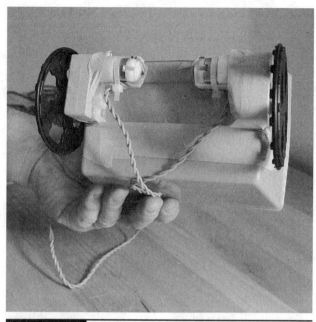
Figure 9-45 Motor wire management.

Step 3: Wire Management and Craft Stick Casters

The Hummingbird board will eventually end up in the container. Run the wires up the opposite side of the container, and hold them in place with a couple of pieces of tape. Many two-wheeled robot kits use a metal or plastic caster to keep the robot balanced. The round edge of a craft stick works perfectly for this task. Turn the container so that you are facing the wheel, and attach a large craft stick to the side of the container with some tape, as in Figure 9-46. Your container will now rest at a slight angle. Repeat the step for the opposite side. You can use tape or hot glue to hold the sticks in place.

Step 4: Blinkers and Brake Lights

Find the orange LED with orange and black wires and the yellow LED with yellow and black wires. These two LEDs will be used as blinkers. Tape the orange LED on the right side of the inside of the container with about 1 inch sticking out above the edge. Do the same thing on the opposite corner with the yellow LED. Now

Figure 9-46   Both casters attached.

locate the RGB LED. We will use this LED to denote whether the robot is moving forward, backward, braking, or in a state of rest. Tape the RGB LED in the middle of the interior backside of the container at the same height as the other LEDs (Figure 9-47). Wire management is not

Figure 9-47   Blinkers and brake lights in position.

necessary at this point; simply lay the wires over the front edge of the container and get ready to connect them.

Step 5: Connect the Motors to Ports

Place the Hummingbird board on the table with the connector side facing up. Find the connection ports marked "Motor 1" and "Motor 2." You may notice that there are two terminal locations for positive and negative wires, but the wires coming from the motors are both the same color. Which lead you put in positive and which in negative will determine whether the motor rotates in a clockwise or counterclockwise direction. Later you may need to switch these wires or reverse the value in the code to make them both rotate in the same direction. Start with the motor on the right side. Press down on the terminal tab with the orange terminal tool from the kit. While pressing down on the tab, push the metal of the wire in at an angle (Figure 9-48). When the exposed metal end of the wire is in place, release the tab on the terminal, and pull lightly on the wire to make

Figure 9-48   Connecting motor wires.

sure that it is secure. Now do the same for the other wire coming from the motor. You will repeat this process for the motor on the left side of the robot, placing the wires in motor port 2.

Step 6: Connect the LEDs to Ports

The tricolor LED will connect to LED port 1. The tricolor port has terminals labeled R for red, G for green, B for Blue, and – for negative. The wires on the LED should be connected to the corresponding colors. The black wire should be connected to the negative terminal.

There are four LED ports on the Hummingbird Duo board, each with a positive and negative terminal. The LED color is denoted by the color of wire coming from the positive lead. Connect the orange LED to LED port 1. Use the positive terminal for the orange wire and the negative terminal for the black wire. Repeat the process for the yellow LED, and connect it to port 2 (Figure 9-49).

Figure 9-49 Fully wired and powered.

Step 7: Get Snap!

You have several options for software to program your Hummingbird robot. For this project, we are going to use Snap! Snap! is a drag-and-drop descendant of Scratch that looks very similar but has added motion blocks for controlling the motors, servos, sensors, and LEDs in the Hummingbird Kit. Follow the directions for installing Snap! and BirdBrain Robot Server on your device. Create a Snap! account so that you can save your work. Before you begin programming, plug in the USB cable and the power supply for the Hummingbird Duo board.

Step 8: Brake and Forward Motion Script

To begin, you need to determine whether the motors roll the same way. You may remember that in the art bot project if you reversed the leads on the dc motor, it would run in reverse. The geared motors in the Hummingbird Kit work the same way. The leads are not denoted, so you may have plugged the battery in so that it runs clockwise or counterclockwise. We can determine which direction motor 1 and motor 2 will run with a few lines of code.

Select and drag the "When space key is pressed" block from the "Control" menu into the scripts window (Figure 9-50). Click the "Motion" menu, and then drag and attach two "Hummingbird motor" code blocks to the "When space key is pressed" code block. You will notice that the "Hummingbird motor" code block has places for two boxes to input values. The first box represents the motor port number into which the motor is plugged. The second box represents the amount of motor speed, which can range from – (negative)100 % for full reverse to 100 % for full-speed forward (Figure 9-51).

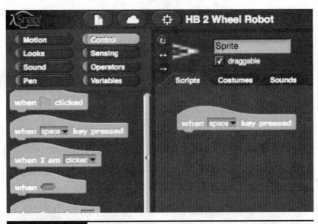

Figure 9-50 Snap! when space key is pressed.

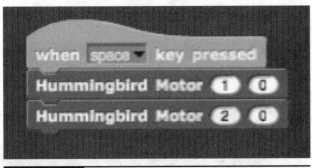

Figure 9-51 Snap! "Hummingbird motor" code block added to script.

Before testing the motors, we need to write a code to stop the motors. Leave the key option set to "Space" in the "When ___ key is pressed" block for both Hummingbird motors. The default port for the "Hummingbird motor" is port 1, which is where our right motor is connected, so there is no need to change the value. In the second block, change the port value to 2 to reflect the port value of the second motor. In the speed box, input the 0 for both motors. Now, when the spacebar is pressed, the speed value of 0 will turn off the motors in ports 1 and 2 (see Figure 9-51).

Step 9: Forward Motion and Motor Direction

Right-click on the script for the brake, and select "duplicate," as in Figure 9-52. Move this script

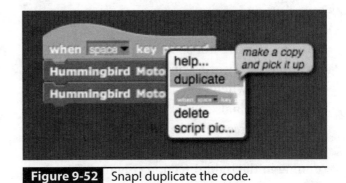

Figure 9-52 Snap! duplicate the code.

over so that you can easily view it. Click on the word "space" in the "When ___ key is pressed" block and change it to "up arrow." You will leave the first value that represents the motor ports the same for both "Hummingbird motor" blocks. However, place a 50 in the second box for speed for both "Hummingbird motor" blocks. This will set both motors to turn at 50 percent speed in a clockwise direction (Figure 9-53).

Lift the robot off the table, and press the up arrow. Watch to see if both motors are spinning in a direction that will cause the robot to travel forward. If they are spinning in opposite directions, note which side is not spinning in the right direction. Press the spacebar to stop them, and place the robot on the surface. If both motors make the robot go forward, skip to step 11.

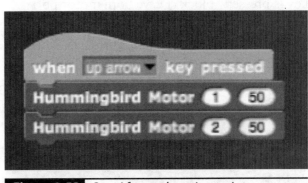

Figure 9-53 Snap! forward motion script.

Step 10: Debug

If the motors are not spinning in the same direction or forward, you have two options. One option is to switch the wires from the motor in the port terminals to make the motor turn in the reverse direction. The second is to change the speed value to –50 for the motor that is not running in the right direction (Figure 9-54). Both options have the same effect, but it may cause confusion to remember to reverse the value for one motor. We recommend that you switch the wires because our examples reflect motors that run in the same direction.

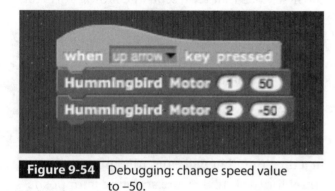

Figure 9-54 Debugging: change speed value to –50.

Step 11: Reverse

Right-click on the forward-motion script, and choose "duplicate." Switch the key option from "up arrow" to "down arrow" in the "When __ key is pressed" block. Leave both motor port values the same, but input –50 for the speed value of both motors, as in Figure 9-55. This should send both motors in reverse when the down arrow key is pressed.

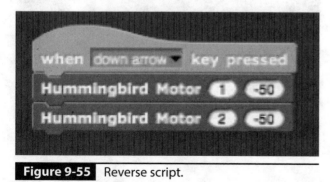

Figure 9-55 Reverse script.

Step 12: Turn Right and Left Scripts

Right-click on the reverse-motion script, and choose "duplicate." Switch the key option from "down arrow" to "right arrow" in the "When __ key is pressed" block. Leave both motor port values the same. To make the bot turn to the right, we will actually turn the right motor off. The right motor is connected to motor port 1, so set the speed value to 0 for motor 1. Motor 2, or the left motor, should have a speed value of 50 (see program in Figure 9-56). This will cause the left motor to keep rotating and spin the bot to the right.

Figure 9-56 Snap! right turn.

To turn left, we will repeat the process by duplicating the previous script and selecting the "left arrow" in the "When __ key is pressed" block. The right motor, motor 1, will have a value of 50 for speed, and the left motor will brake with a value of 0 for speed (Figure 9-57). Test your bot out, and make adjustments to speed, if necessary.

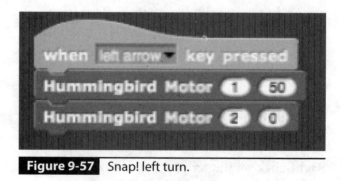

Figure 9-57 Snap! left turn.

Step 13: Blinkers

No civilized robot would ever dream of making a turn without the proper signal! Select the "Control" block menu. Locate the "Repeat 10" code block, and add it to both the right and left motion scripts you have created. Next, select the "Looks" block menu, and click and drag the "Hummingbird LED" block into the "Repeat 10" block. The first value in the "Hummingbird LED" block represents the port the LED is connected to. We connected the orange LED on the right side into port 1, so leave the default value of 1 for the right-turn script. The second value represents the brightness of the LED. Change this value to 80. The yellow LED on the left side of the robot is plugged into LED port 2. Change the first value of the "Hummingbird LED" block to 2 to reflect this, and set the brightness value to 80 (Figure 9-58).

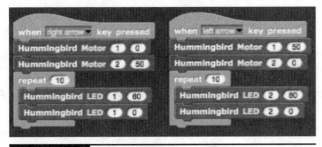

Figure 9-58   LED on and off.

To make the LED blink, we need to turn it off and on with a pause in between each time. To do this, add another "Hummingbird LED" block below the first "Hummingbird LED" block in both the right and left scripts. Set the port value to reflect LED 1 or 2, and change the brightness value to 0. This will make the LED turn off. To keep the LED from switching on and off too quickly, add a "wait" code block from the "Control" menu after each "Hummingbird LED" block. Set the value to 0.2 second in each of the "wait" code blocks (Figure 9-59). Test your code by pressing on the right and left

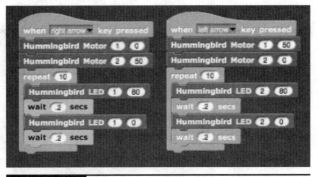

Figure 9-59   Waiting to blink.

arrows. The LED should blink on and off now for each side.

Step 14: Forward, Reverse, Rest, and RGB LED

The LED in the center of the robot is capable of changing colors. Locate the "Hummingbird tri-LED" block from the "Looks" menu, and drag a copy of it to the "forward," "reverse," and "brake" scripts. The first value in the "Hummingbird tri-LED" block represents the port numbers. We connected the tri-LED to tri-LED port 1, so we will leave the default value of 1 in place. The next values labeled "R" for red, "G" for green, and "B" for Blue control the brightness of that color in the mix. The values can range from 0 to 100.

For the forward script, leave the default value of R 0, G 100, and B 0. This will make the LED turn green when the robot is in forward motion. When the robot is in reverse, we want a red warning light, so in the reverse script input the values R 100, G 0, and B 0. When the robot stops, we want the LED to turn purple. Set the brightness values to R 50, G 0, and B 50. Remember, our settings in Figure 9-60 are just a guide. Feel free to customize your robot with more LEDs and color choices!

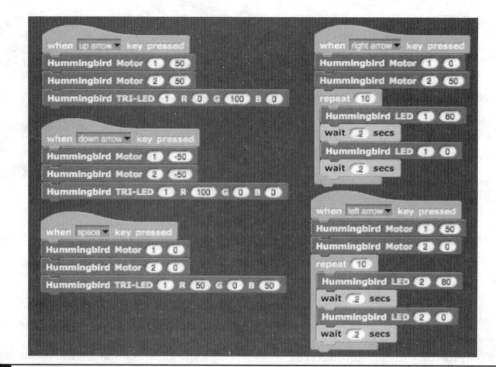

Figure 9-60 Complete Snap! Program.

Step 15: Improve Mobility

You have probably noticed that the plug-in power adapter and mess of cables coming from the robot to the Hummingbird Duo board is somewhat limiting. We recommend that you purchase a battery power source or follow the instructions to make an auxiliary power source on the Hummingbird website FAQ webpage. While using Snap!, you will still be limited to the USB cable for distance, but your wires can be tucked into the container you are using, and the Duo board can be placed inside or mounted as in Figure 9-61.

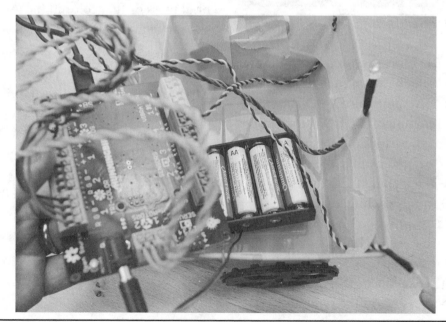

Figure 9-61 Battery pack.

Challenges

- Try using a different software to program the Hummingbird Kit. How can you use the sensors or other items to make your bot even more amazing?

- Can you use Ardublock to program your Hummingbird? Can you upload a dancing program? A figure eight?

- Can you use "if/then" statements to program your Hummingbird robot to do certain actions if the sensors sense something?

"App-Smashing Programming" Challenge

Mastering programming in one app is an accomplishment, but the ability to re-create the same program in another app proves that you are a coding master! Try some of the same challenges in this chapter but with a different app.

Let us know how you do by tweeting to @gravescolleen or @gravesdotaaron. Or upload pictures with the custom #bigmakerbook hashtag to be featured on our community page!

littleBits

LITTLEBITS IS AN amazing prototyping tool. Get an idea, and test it out immediately by combining littleBits with cardboard and other stuff you find around your house! The projects in this chapter could be adapted to fit all sorts of other projects. Plus, one of the best things about littleBits is that you can use many different Bits to complete the same functions. What Bits will you use, and what will you do with your invention?

Project 41:	Robot Arms and Moving Gates
Project 42:	Flashing Rainbow Lamp or Tunnel
Project 43:	Classic Miniature Golf Windmill

Chapter 10 Challenge

Prototyping challenge: How does ___ work and can I make my own?

Project 41: Robot Arms and Moving Gates

Cost: Free–$$–$$$

Make time: 15–30 minutes

Supplies:

Materials	Description	Source
Cardboard	Cardboard boxes	Recycling bin
Sticky stuff	Duct tape, masking tape, zip ties	Hardware store / Craft store
Tools	Box cutter, craft knife, pencil, Phillips screwdriver	Hardware store
littleBits	Servo Bit (o14), pulse Bit (i16), power Bit (p1) with cable and 9-V battery	littleBits.cc
littleBits (optional)	Motion trigger Bit (i18)	littleBits.cc

Servo motors are a great way to make something move. You could combine this project with a Makey Makey and make a "Don't Eat My Cookies" robot arm, or you could start thinking about making a full robot. With this project, you'll have some moving robot arms, so once you get done, start thinking about how you will invent the rest of your robot.

Step 1: Arm or Gate

Cut a rectangular shape of cardboard about
3 × 12 inches. Measure in about 2 inches from
the short side, and place a mark near the middle.
Place the cardboard over a roll of tape, and
slowly turn and press a screwdriver or pencil
through the mark so that you have a hole
(Figure 10-1). This hole will allow you to access
the screw that holds the servo arm on so that you
can make adjustments to the swing, if needed.

Figure 10-1   Hole in the arm.

Step 2: Servo Rotation Range

For this build, we are going to attach the servo
to the side of a box and make our arm swing up
and then down parallel to the surface on which
the box is resting. Place the servo arm on the
side of the box, and rotate it all the way to the
right. This is a way to determine where you will
need to position the servo arm and to adjust it
to fit your needs. When the servo is all the way
to the right, it needs to be parallel to the ground.
Remove the screw, and adjust the servo arm at
this time. Make sure that the mode on the servo
motor is set to swing and secure the plastic servo
arm in place with the screw (Figure 10-2).

Figure 10-2   Servo in correct position.

Step 3: Attach Robot Arm or Gate to Servo

Center the servo over the hole in the robot arm
so that you will be able to access the screw later
for adjustments, as in Figure 10-3. Use a couple
of strips of tape to hold the cardboard to the
servo. Tightly wrap over these with tape. It's
important that the servo arm is firmly held in
place to the robot arm.

Step 4: Servo Position on the Box

Measure 3 inches up from the bottom of the
box, and place a mark. The output spline is the
part of the servo on which the servo arm slides.
Place the servo on the side of the box, and center
the output spline on the mark. With the servo
in position, mark the top and bottom of the
rectangular body of the servo on the side of the
box. To punch out holes, twist a screwdriver
into the box based on your marks. Attach the
servo with a zip tie as tightly as possible, as in
Figure 10-4.

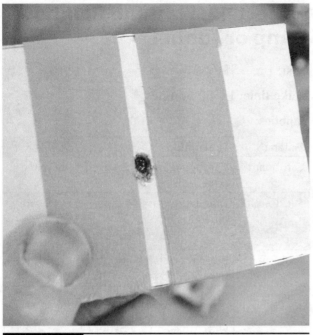

Figure 10-3 Servo taped in position with access to hole.

Figure 10-4 Servo in position and zip tied.

Step 5: Attach the Circuit

Think about whether you want to control the arm/gate with a motion sensor, button, or even a wireless remote! It may be helpful to test your arm with a button before automating the circuit with a motion sensor and pulse. Starting on the left with your power supply and power Bit, attach a button and a wire Bit (w1) to the servo (o11) you are testing (Figure 10-5). If you plan on automating the robot, attach the pulse Bit (i16), and adjust the speed of the pulse to as

Figure 10-5 Unattached basic circuit.

Figure 10-6   Servo gate at work.

slow as possible. If you want to add a motion trigger, place it directly after the power and before the pulse; then turn on your circuit and test it out! Once you've completed testing your circuit and debugged for problems, secure the littleBits circuit to the box. Place it on a Sphero track, use it as a prank, or use it as a robot arm! See the finished project in Figure 10-6.

Challenges

- How can you build another arm that is controlled with this arm?

- What creative way can you decorate the gate? Hammer of Thor, anyone?

- For what other functions could you use this karate-chop motion?

- How could you incorporate the servo and a wireless transmitter to prank someone?

Project 42: Flashing Rainbow Lamp or Tunnel

Cost: Free–$$–$$$

Make time: 15–30 minutes

Supplies:

Materials	Description	Source
Cardboard	Cardboard boxes, tissue paper, foil	Recycling
Sticky stuff	Duct tape, masking tape, glue, zip ties	Hardware store Craft store
Tools	Box cutter, craft knife, hot glue gun	Craft store
littleBits	RGB Bit (o3), pulse Bit (i16), wire Bit (w1), power Bit (p1) with cable and 9-V battery	littleBits.cc

Step 1: Marky Mark and the Slats

For this project, you will need a long, narrow box. The box we used for this build was about 8 × 6 × 18 inches. Place a mark every inch on both sides of the top of the box. Connect the marks, and place an X on on every other strip. Turn the box on its side, and mark a line down the horizontal center of the box. We placed our line at half the height of the box at 3 inches. Make a mark every inch on this line, and then connect the existing lines from the top to the mark on the side, as shown in Figure 10-7. We will cut out a strip every inch, so you may want to mark the pieces with an X so that you won't get confused at what strip to cut and what strip to keep.

Step 2: Cut the Slats and Possibly Doors

Use a box cutter to cut out every other strip of cardboard (Figure 10-8). Avoid cutting the strips next to the end of the box so that you have a 1- or 2-inch border depending on the length of your box and how the pattern works out. See finished slats in Figure 10-9.

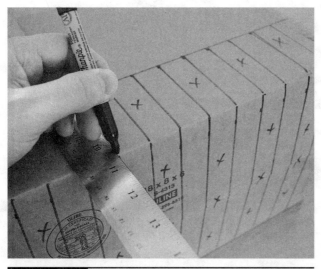

Figure 10-7 Marking slats on top and sides and marking with an X.

Figure 10-9 Finished slats.

If you want to make a lamp, leave the sides in place, and simply close the open end after you mount the littleBits inside.

Step 3: Paint and Diffuse the Light

At this point, you can lightly spray your cardboard with spray paint to make a superswanky lamp. Use only a very light coat so that the cardboard does not warp. If you like the industrial look of cardboard or have a cool box, feel free to just leave it!

You can add tissue paper to help diffuse the light by spreading it out evenly (Figure 10-10). It also allows you to cover up the workings of the lamp. If you don't have any RGB LEDs, you can use colored tissue, and it will look just as cool. Turn the box over, and cut a rectangle of tissue paper about 2 inches larger than the cut-out section. If you are using colored strips of tissue paper, just make sure that the strips are large enough to be glued to the sides of the slat. Use an even coating of glue to hold the tissue paper in place. Cover the rest of the interior of the box with aluminum foil using glue or tape as an adhesive. The aluminum foil will reflect the

Figure 10-8 Cutting the slats.

At this point, you need to decide whether you want to make a lamp or if you want to make a tunnel for robots or toys. If you are making a tunnel, mark a door for the robot to enter and exit on the short side of the box. The size of the tunnel entrance will vary depending on the size of the toy or robot you want to use (Figure 10-9). The open side of the box will become the bottom of your tunnel. Fold the flaps on the outside, and you will have an easy way to tape your tunnel in place, as shown in Figure 10-9.

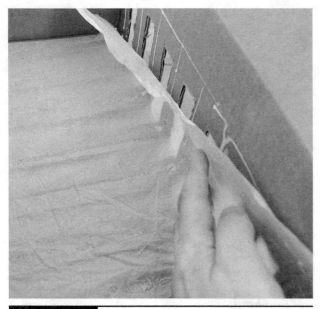

Figure 10-10   Glue and place tissue paper.

Figure 10-11   Circuit setup for inside tunnel.

Figure 10-12   Circuit setup for outside tunnel.

light and make your lamp brighter. Wait a few minutes for the glue to dry, and start working on your littleBits circuit.

Step 4: The Circuit

The littleBits circuit uses the power Bit (p1), four to eight RGB LED Bits (o3), and a pulse Bit (i16). If you don't have eight RGB LED bits, you can use LED Bit (o1), the bright LED Bit (o14), or the bar-graph Bit (o9) with colored tissue paper as substitutes. If you have the RGB Bits use the r, g, and b adjustments to mix a rainbow or whatever color combination pleases you. Attach the pulse Bit (i16) before the LED Bits, if you want your tunnel to flash. Make sure that you assemble your colors in the order you want to create a rainbow (Figure 10-11). Slide your circuit inside the box, and test it. Try different positions, and mark the spots that best light up your tunnel/lamp (Figure 10-12).

Step 5: Bits in a Channel

You can use the littleBits adhesive shoes to mount your circuit or use a slice of cardboard and some rubber bands to create a channel in the box for the circuit to rest on. If you are making a lamp, add wire (w1) to the circuit so that you can place the switch on the outside. To create the channel, start by cutting a 2-inch-wide strip of cardboard. Make a mark at ½ inch and at 1½ inches on both short sides of the cardboard strip (see Figure 10-11). Connect the marks, and then use a box cutter to score one side of the cardboard along these lines. Fold the cardboard into a U-shaped channel along these lines. Slide the circuit into the channel about 1 inch away from the end, and mark about 1 inch after where the circuit ends. Use some scissors to slice the upward turns of the U so that you have two flat flaps on the end of the channel (Figure 10-13). Wrap a few rubber bands over the circuit to secure it, and use the flaps to tape the circuit in place (Figure 10-14). Hot glue also works great if you have it available. If you are making a lamp, run the wire out of a corner of the box, and tape it up. It's also possible to use a zip tie to secure the power and input circuit to the box. Just remember not to tighten the zip tie too tightly. After you are finished, power it up, step back, and enjoy the show! See the finished project in Figure 10-15.

Classroom tip: If you are working with a large group, this is a good time to make a couple of lamps and let the participants responsible for mixing colors learn more about RGB color mixing. It is also a great time to

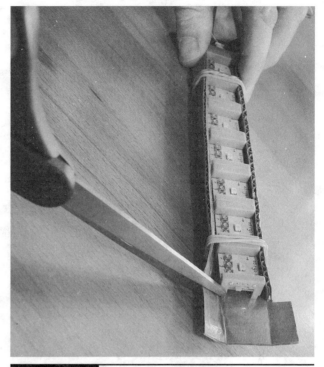

Figure 10-13 Securing littleBits in cardboard channel.

experiment with light diffusion. What happens if you use a darker piece of tissue paper versus white tissue paper? What happens if you use yellow tissue paper? Orange? Blue?

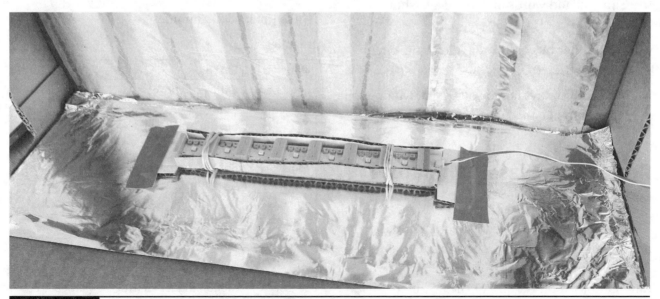

Figure 10-14 Lighting in place inside tunnel.

Figure 10-15 Rainbow tunnel lights.

Challenges

- What other materials could be used to diffuse the light? What happens if you use Styrofoam? Cotton balls?

- Can you add words and make a flashing sign?

- What other bits could you add to the circuit to really trick this light out?

- What other ways could you design windows for a lamp like this?

Project 43: Classic Miniature Golf Windmill

This project is a great decorative item but also an essential piece for any robot obstacle course or miniature golf course. Every build will be different based on the size of the box used. This is a general guide that will teach the basic

steps and allow you to substitute your own measurements and customization. Start with a large box that can stand on the smallest side. The box for this build was 12 × 12 × 20 inches.

Cost: Free–$$–$$$

Make time: 15–30 minutes

Supplies:

Materials	Description	Source
Cardboard	Cardboard boxes	Recycling
Sticky stuff	Duct tape, masking tape, hot glue sticks, zip ties	Hardware store Craft store
Tools	Box cutter, craft knife, hot glue gun	Craft store
littleBits	DC motor Bit (o25), slide dimmer Bit (i5), wire Bit (w1), power Bit (p1) with cable and 9-V battery, MotorMate	littleBits.cc
Sensor (optional)	Motion trigger Bit (i18)	littleBits.cc

Step 1: Make Stock and Windmill Mount

Cut two 22-inch strips that are 3 inches wide. If your box is larger or smaller, the length will vary. As a general rule, double the height of the box, and make your strips to this height. On a traditional windmill, these pieces are called *stock* and hold the giant sails or sweeps. Place a line at the center of each cardboard strip and a parallel line on both sides at 1½ inches away, as in Figure 10-16? Center the strips, and use hot glue to hold them together. Avoid placing glue in the center. Use a screwdriver or pencil to twist a hole in center. To avoid gouging your work surface, place the center of the cross shape over a roll of tape, as shown in Figure 10-16. Use a standard pencil to enlarge the hole and push the littleBits MotorMate through the center (Figure 10-17). Once the MotorMate is through, push a craft stick into the large groove on the MotorMate, and use hot glue to secure it (Figure 10-18). Later, you can add a drop of hot glue to the end of the motor shaft and attach the MotorMate, but for now, we need to be able to slide it on and off.

Figure 10-17 MotorMate test fit.

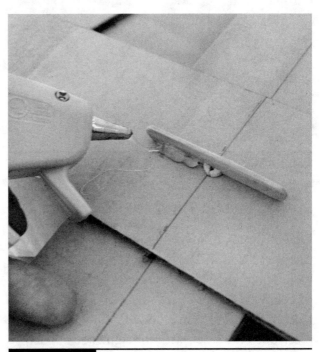

Figure 10-18 Secure the craft stick.

Figure 10-16 Hole punching with a pencil.

Step 2: Mount the Motor

Find the center of the top of the box, and place a mark. Center the dc motor Bit (o25) with the shaft hanging over the edge, and place a mark in the center of the left and right sides. Poke a hole in these two spots with a screwdriver. Use a zip tie to securely fasten the motor to the top of the box (Figure 10-19). If you are using the old-style motor bit, use some slices of cardboard under the bit to keep from bending the bit and avoid fastening it too tight.

Figure 10-20 Testing and marking sweep height on stock.

Figure 10-19 Dc motor secured with zip tie.

Step 3: Sweep Time

Now it is time to make the sweeps that attach to the stock. You will need to cut four 6- × 8-inch rectangular shapes. Slide the cross-shaped stock piece on the motor shaft, and turn it so that the stock is pointing straight down. Hold a sweep piece onto the stock with the right side aligned with stock, and rotate it slightly. Adjust the height so that the sweep will clear the ground, and mark the height on the stock. Mark the height for all four pieces; then use hot glue to secure them in place (Figure 10-20). Make sure that the right side of the sweep is aligned

with the stock piece so that you get the classic windmill look.

Step 4: What About the Doors?

If you want to use this windmill as an obstacle, you will need to cut doors on the front and back. Remove the windmill sweep structure and circuit before cutting. The width of the doors should be 6 inches, and you can determine the height of the door by measuring the robot or object you want to pass through the windmill (Figure 10-21).

Step 5: Motor Circuit and Motion Detection

Connect a battery to the power Bit (p1), and attach a slide dimmer Bit (i5) to the dc motor (o25). Then turn the dimmer all the way down.

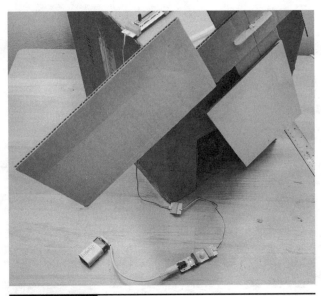

Figure 10-21　Cutting the door.

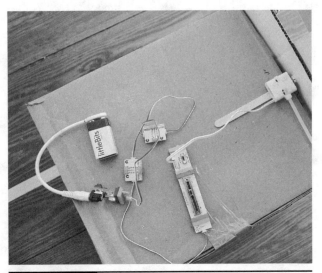

Figure 10-22　Windmill circuit with dimmer.

Set the switch on the dc motor to "CW" for clockwise or "CCW" for counterclockwise. Turn the circuit on, and adjust the dimmer to the perfect speed (see circuit in Figure 10-22).

If you don't want to leave your windmill going all the time, you may want to add a motion sensor to your circuit. If you use the motion sensor, you'll need to place some distance between the rotating blades and the sensor, so you'll need several wires (w1) (shown earlier in Figure 10-21). Position the motion sensor at the beginning of the circuit, as if it were a button. Place the sensor in a spot where it will be triggered by an approaching person or robot. Use the wires to connect the sensor to the motor. To build a channel to hold your circuit, revisit step 5 or Project 48. Another option is to use the adhesive mounting shoes or tape a mounting board to the box.

Challenges

- What other bits can you use to make the windmill an even more amazing attraction?

- Could you use your Arduino knowledge to make an "if/then" statement that activates the windmill with a motion sensor?

- What other ways could you hack this windmill?

"How Does ___ Work? And Can I Make My Own?" Challenge

Because littleBits is the ultimate prototyping tool, could you use it to invent versions of everyday things? Or could you use it to make everyday things even better? Go to the littleBits projects page to see other amazing inventions with littleBits, and join the Bitster community to get inspired on a daily basis!

Share your projects with us using our custom hashtag #bigmakerbook! And don't forget to tag @littleBits!

3D Printing

IN THIS CHAPTER we explore some projects that use three-dimensional (3D) printing and design to enhance other makerspace resources. You don't need to own or even have access to a 3D printer to learn about 3D modeling and design. Check with your local library about printing availability and cost. Tinkercad also has several 3D printing partners that can print your 3D designs. Chances are that for just a few bucks, you can produce your own prototypes in no time!

Project 44:	Designing for Makey Makey—Earth Bracelet
Project 45:	Designing for littleBits—Wheels and Pulleys for a DC Motor
Project 46:	DIY Phonograph Top
Project 47:	Sphero Paddles

Chapter 11 Challenge

"3D printing" challenge.

Project 44: Designing for Makey Makey—Earth Bracelet

We love Makey Makey, but sometimes we wished we had an easy way to ground players to earth fashionably. In this project, we learn how to 3D design a basic earth bracelet that incorporates a little copper tape to keep even the wildest banana pianist completing circuits (Figure 11-1).

Cost: $–$$

Make time: 45 minutes–2 hours

Supplies:

Materials	Description	Source
Measurement tools	Ruler Marker	Craft store
3D modeling software	Tinkercad is a browser-based 3D modeling and design software. You will need access to the Internet and a browser and computer that meet Tinkercad's requirements	Tinkercad.com
3D printer and filament	These projects were made on a variety of printers available through public libraries.	Public library
Conductive tape	Copper tape or soft conductive tape in Makey Makey Inventor's Booster Kit	SparkFun Joylabz.com
Makey Makey Invention Kit	Makey Makey, alligator test leads, USB cable, and wires	Joylabz.com

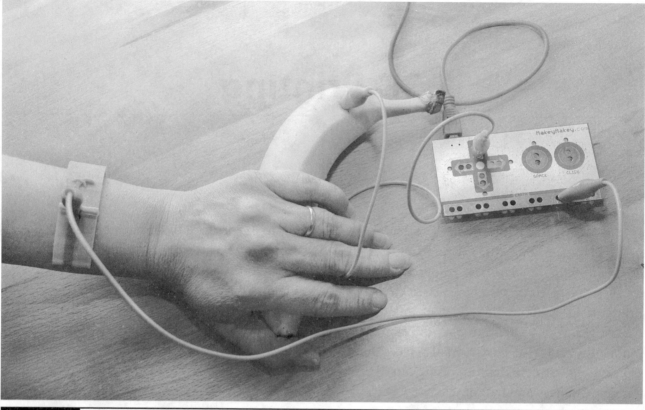

Figure 11-1 Makey Makey earth bracelet.

Step 1: Measure Your Wrist

To get an idea of how big to make your bracelet, you need to take some measurements of your wrist. Use a ruler to estimate the width across your wrist. Turn your wrist sideways, and measure it to determine the depth you will need to make the bracelet shape. The measurements will be different for everyone. Our example features dimensions for a teen or average adult.

Step 2: Size the Bracelet

Throughout this project, you may find it helpful to change the measurement on the snap grid to make aligning shapes easier. Start by clicking the "Edit Grid" button in the lower right corner and changing the unit to inches, as shown in Figure 11-2. Add ¼ inch to the dimensions you took earlier, and resize the cylinder to this width and depth. For our example, we measured a wrist

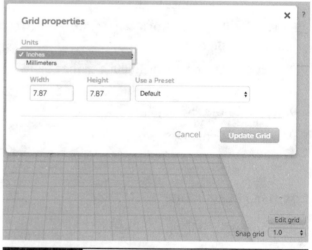

Figure 11-2 Edit grid.

that was $2\frac{5}{8} \times 1\frac{3}{4}$ inches, so we will adjust the size of the cylinder to $2\frac{7}{8} \times 2$ inches. Drag a cylinder from the "Geometric Shapes" menu to the workplane. Use the white box on the top to adjust the height of the cylinder to ¾ inch. Use

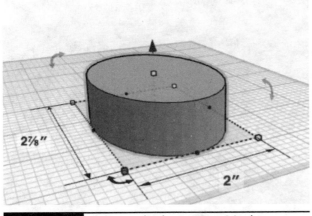

Figure 11-3 Adjust cylinder to 2⅞ × 2 inches.

the white box in the corner to adjust the size to 2⅞ × 2 inches (Figure 11-3).

Now we need to create a hole in the cylinder to make it into a bracelet. Click on the "Hole Shape" menu, and drag the cylinder-shaped hole to the workplane. Click the "Ruler" tool, and place it on the lower-right side of the cylinder hole shape. Resize the hole to your wrist measurements. The dimensions for our example are 2⅝ × 1¾ inches (Figure 11-4). It's not necessary to resize the height because the default is set to 1 inch. Center the hole in the cylinder so that your bracelet is now ⅛ inch thick. To align the hole perfectly, select both shapes, and then click on the "Adjust" menu in the upper right

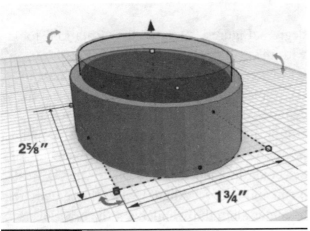

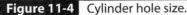

Figure 11-4 Cylinder hole size.

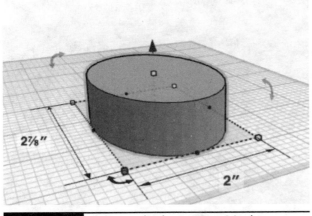

Figure 11-5 Cylinder hole alignment.

of the screen and choose "Align." The "Align" tool creates several marks that correspond with left, center, and right alignment, as well as top, middle, and bottom (Figure 11-5). Click on the center align mark on the side and front of the cylinder to center the cylinder hole. Now that both shapes are aligned, click "Group" next to the "Adjust" tool to group your two shapes and cut a hole in your bracelet.

Step 3: How to Get into It

We have a bracelet shape now but no way to get our arm into it. To remedy this, select the box hole, and resize it ½ inch smaller than the width of the side wrist measurement. In our example, the side of the wrist measured 1¾ inches, so we will make our box hole 1¼ × 1¼ inches and place it though the bracelet on the short end (Figure 11-6). Once printed, the plastic bracelet should flex enough to allow the user to slide it on while it still fits snugly. Use the "Align" tool to center the box hole on the short edge. Wait to group the two shapes until after the next step. We need the whole cylinder shape in place to center some other components.

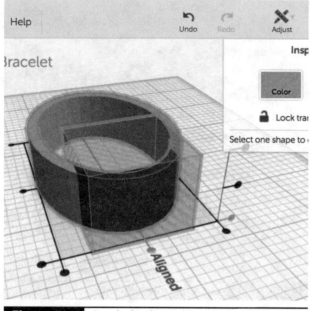

Figure 11-6   Box hole aligned.

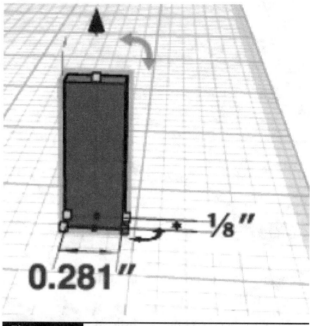

Figure 11-7   Box resized.

Step 4: Alligator Clip Terminal

Next, let's create an easy place to clip the alligator clip test lead from the Makey Makey that adds a cool decorative look to the bracelet. Rotate the workplane so that one of the long sides of the bracelet is facing you. Select a box from the "Geometric Shapes" menu, and drag it to the workplane. Use the white dot on top to resize the height to ¾ inch. Adjust the box width to ⅛ inch and a depth of 0.281 inch (Figure 11-7). You may need to change the snap grid to ¹⁄₆₄ inch to achieve this measurement or use the "Ruler" tool. Position the rectangle in the center of the bracelet, and align the furthest edge from you with the inside edge of the bracelet. Use the "Align" tool to place the rectangle in the center, as shown in Figure 11-8. Select all the shapes, and click the "Group" tool.

Now that we have a rectangular shape for the alligator clip to grasp onto, we can add some holes to make sure that it is really secure. Create a ⅛- × ⅛-inch cylinder hole. There is no need to worry about the height, but you need to click on the rotation arrows and turn the cylinder 90

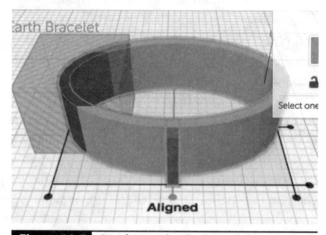

Figure 11-8   Box for test leads aligned.

degrees. Figure 11-9 shows the "Rotation" tool with the cylinder hole in the proper position for placing it into the bracelet. Place it next to the test lead box, as in Figure 11-10. Select the cylinder, click the "Edit" menu, and choose "duplicate." Change the snap grid to 1/64 inch to position a cylinder on each side of the rectangle. Select all the shapes, and then click the "Adjust" menu and choose "Align." Select the center circle to center align the cylinders on the bracelet. Select all the shapes, and press the "Group" button.

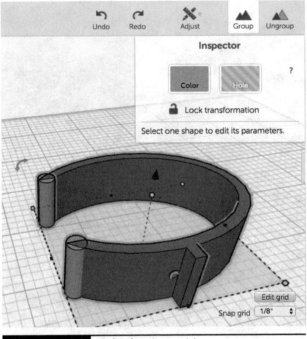

Figure 11-9 Rotating ⅛-inch hole.

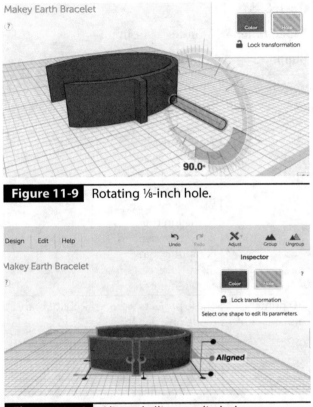

Figure 11-10 Aligned alligator clip holes.

Step 5: Round the Edges

The opening is complete, but we need to clean up the sharp edges. Drag a cylinder shape to the workplane from the "Geometric Shapes" menu (Figure 11-11). Use the white dot in the corner or "Ruler" tool to resize the cylinder width and depth to 0.219 inch. Click on the

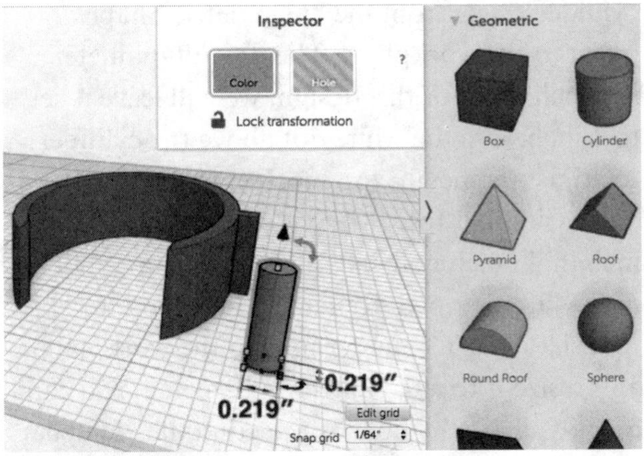

Figure 11-11 Cylinder resized.

Figure 11-12 Cylinders in position.

white dot on the top of the cylinder to adjust the height to ¾ inch. When you have the cylinder resized, click on the "Edit" menu, and choose "duplicate." Position the cylinders at the open ends of the bracelet to cover the sharp edges (Figure 11-12). Select the bracelet and cylinders, and click "Group" to join them into one shape. At this time you can customize the bracelet with other items if you want, but the basic design is complete.

Step 6: Download, Print, Conduct, then Connect

Click on the "Design" menu and select "Download for 3D Printing." Choose either STL or OBJ for the file type based on the requirements of your 3D printer. After you have printed the project, add a strip of copper or conductive tape down the inside center of the bracelet that ends near the alligator clip terminal, as shown in Figure 11-13. Connect an alligator clip to earth on the Makey Makey and to the bracelet, and complete a circuit!

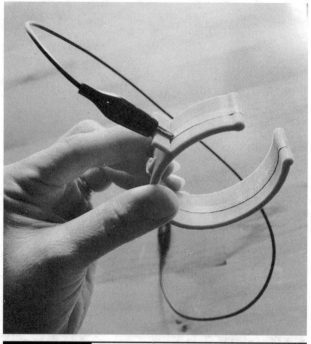

Figure 11-13 Conductive tape and final bracelet.

Challenges

- What other ways can you use 3D printed objects to create with a Makey Makey?

- Could you design a ring for your earth connection?

- What about a friendship bracelet for two players?

Project 45: Designing for littleBits—Wheels and Pulleys for a DC Motor

The littleBit MotorMate is an amazing tool for adding Lego axles and wedging craft sticks and cardboard into your projects. However, sometimes you may have a need to make your own wheel or pulley for a project. This project shows how to make a basic pulley and wheel that you can customize to fit your needs.

Cost: $–$$

Make time: 45 minutes–2 hours

Supplies:

Materials	Description	Source
Measurement tools	Ruler	Craft store
3D modeling software	Tinkercad is a browser-based 3D modeling and design software. You will need access to the Internet and a browser and computer that meet Tinkercad's requirements.	Tinkercad.com
3D printer and filament	These projects were made on a variety of printers available through public libraries	Public library
littleBits	Dc motor Bit (o25), power Bit (p1) with cable and 9-V battery, MotorMate	littleBits.cc

Step 1: Basic Shape

Begin this project by clicking the "Edit Grid" button in the lower-left corner and selecting "inches." Remember that during the project it may be necessary to adjust the snap grid to re-create the dimensions and positions described. To create the main body of the pulley, drag a cylinder shape from the "Geometric Shapes" menu to the workplane. The default diameter is 1 inch, and for this design, we will leave it set to 1 inch. Use the white dot above the cylinder to adjust the height to ½ inch (Figure 11-14). The littleBits MotorMate is about 0.32 to 0.34 inch in diameter when you wedge a craft stick in the large groove. To make a hole close to that size, drag a cylinder hole to the workplane and resize it to a 0.34 inch in diameter. You can quickly resize the hole by selecting the "Helpers"

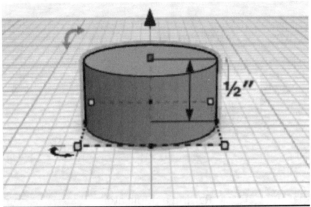

Figure 11-14 Cylinder height.

menu and clicking on the ruler. Click on the white dot in the lower-left corner to reveal the measurements for the shape. Select the width and depth measurement, and enter 0.33 (Figure 11-15). After you have adjusted the diameter of the cylinder, position the hole close to the center of the main body. Click the "Adjust" menu, and select "Align." Click the center alignment marks on the front and side of the shape to put it in the exact center (Figure 11-16). Select both, and click "Group."

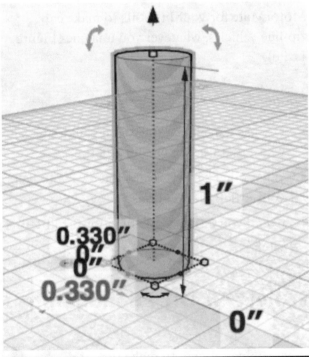

Figure 11-15 Resizing cylinder hole with "Ruler" tool.

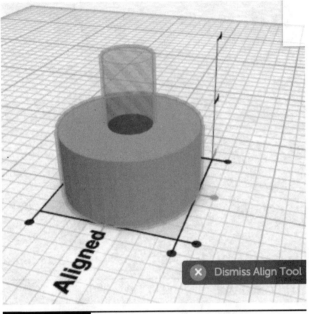

Figure 11-16 Cylinder hole alignment.

Step 2: Center Groove

The groove in the MotorMate is about 0.060 inch wide and 0.30 inch deep. The groove in the MotorMate is meant to flex so that it can hold things tightly. We are going to add a thin box shape to our pulley that will fit into the groove. You will need to click the "Helpers" menu and use the "Ruler" tool to help get these precise measurements. Drag a box shape to the workplane, and adjust the height to 0.3 inch. Set the width to 0.059 inch and the length to ⅞ inch. Position the thin box in the center of the hole. Select both shapes, and then click the "Adjust" menu, and select "Align." Click on the center alignment marks on the front and side to center the box (Figure 11-17). Select both shapes, and click "Group."

Step 3: Add a Belt Groove to the Pulley

If you are making a wheel, this is a good place to stop, but if you are creating a pulley, you'll need to make a belt groove. Rubber bands make great drive belts as long as they have a groove to hold

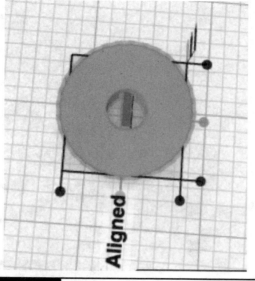

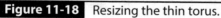

Figure 11-17 Aligning box in center.

them into place. Select the "Thin torus" shape from the "Geometric Shapes" menu and drag it the workplane. Adjust the height to $1/8$ inch and the diameter to $1\frac{1}{8}$ inch (Figure 11-18). Roughly center the torus on the cylinder, and click "Adjust" and select "Align." Click the centering marks provided by the "Align" tool to correctly position the torus. Duplicate the shape and raise the duplicate so that it aligns with the top edge of the cylinder, as in Figure 11-19.

Step 4: Complete the Shape and Print

Select all the shapes and click "Group" to form the complete shape. Click the "Design" menu and select "Download for 3D printing." Choose the appropriate file type for your printer. Now you'll be able to put this pulley wheel on a

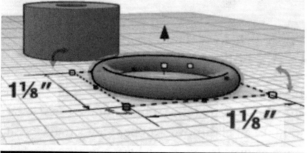

Figure 11-18 Resizing the thin torus.

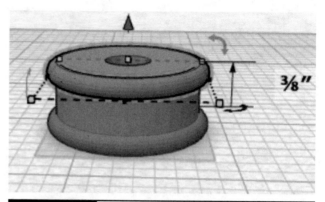

Figure 11-19 Raising a duplicate.

Figure 11-20 Pulley in action.

MotorMate for your littleBits to make cars, zip-line vehicles, whatever you imagine (Figure 11-20).

Challenges

- O-rings make great tires. Can you design a wheel that uses an O-ring as a tire?
- What other objects could you design to help your littleBits inventions?
- Could you design something for the littleBits servo?

Project 46: DIY Phonograph Top

What can you design to make the DIY phonograph player louder and to play more consistently? We thought about this for a long time and decided that to improve the performance of our phonograph, we needed something to keep us from drawing all over our table and constantly having to sharpen our pencil. We decided to make a top-like shape to fit the pencil into, as shown in Figure 11-21. We hope you come up with more ideas on your own for this one, but here are the steps to make one small improvement.

Cost: $ - $$

Make time: 45 minutes

Supplies:

Materials	Description	Source
Measurement tools	Ruler	Craft store
3D modeling software	Tinkercad is a browser-based 3D modeling and design software. You will need access to the Internet and a browser and computer that meet Tinkercad's requirements.	Tinkercad.com
3D printer and filament	These projects were made on a variety of printers available through public libraries.	Public library
Pencil	Unsharpened standard No. 2 pencil	School supplies
Previous project	DIY record player	Project 23

Step 1: Body

This project uses measurements in inches, so be sure to click the "Edit Grid" button and change the setting to inches. After you update the grid, drag a cylinder shape to the workplane, and resize the diameter to 1½ inches. Use the white dot on top of the cylinder to adjust the height to ⅜ inch.

Figure 11-21 Pencil top improvement.

Step 2: A Pencil Holder

A standard pencil is about 0.281, or ⁹⁄₃₂, inch in diameter. We need to make a hole in the cylinder shape to hold a pencil. Change the snap grid to ¹⁄₁₆ inch. Drag a cylinder hole to the workplane, and resize it to a diameter of 0.313 inch. Click the "Adjust" menu and select the "Align" tool. Use the center alignment marks to position the hole in the middle of the cylinder, as shown in Figure 11-22. Click the "Group" button to combine both the hole and the cylinder.

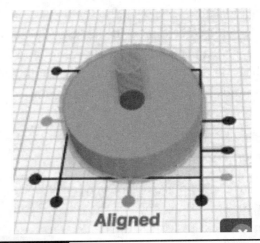

Figure 11-22 Cylinder hole aligned.

Step 3: The Pointed End

Now that we have a place to secure the pencil, it is time to construct the pointed end of the top. Drag the "Cone shape" from the "Geometric Shapes" to the workplane. Use the white dot in the corner to resize the base of the cone to a radius of 1½ inches. Use the white dot at the top of the cone to adjust the height of the cone to ¼ inch. Use the black arrow at the top of the cone to raise it off the workplane ⅜ inch. Position the cone over the cylinder as in Figure 11-23. Select all the shapes and click "Group."

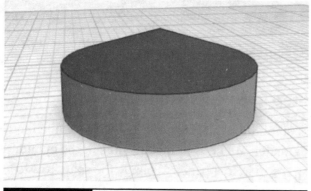

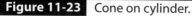

Figure 11-23 Cone on cylinder.

Step 4: Download and Spin

Click the "Design" menu, and then select "Download for 3D printing." Choose "STL" or "OBJ" based on your printer's file preferences. After the print job is over, flip the top so that the pointed side is resting on the table, and place an unsharpened pencil through the record and into the hole on the base. Rotate the record and enjoy the sounds of an easy-to-spin DIY phonograph. (See Figure 11-21.)

Challenges

- This top works great for playing records with a standard hole, but how could you improve it to work with 45 rpm records with a 1½-inch-diameter hole?

- What other objects could you create to make the DIY phonograph play louder and more consistently?

- Could you design a movable arm for your record player? Or a stand for your cone?

Project 47: Sphero Paddles

Sphero is an agile, speedy robot on land, but when you place it in the water, it really slows down. Why not have a robot that can rule both land and sea! In this project, we will design a set of paddle fins that will let Sphero rule the pool, as in Figure 11-24.

Cost: $- $$

Make time: 45 minutes

Supplies:

Materials	Description	Source
3D modeling software	Tinkercad is a browser-based 3D modeling and design software. You will need access to the Internet and a browser and computer that meet Tinkercad's requirements.	Tinkercad.com
3D printer and filament	These projects were made on a variety of printers available through public libraries.	Public library
Waterproof robot	Sphero	Orbotix
Grip buffer	⅛-inch-thick craft foam or packaging foam	Craft store Recycling
Fasteners	Zip ties	Hardware store

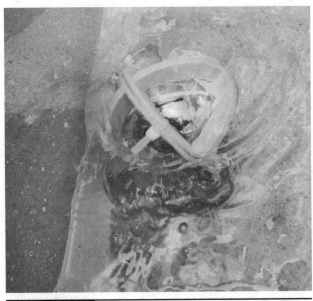

Figure 11-24 Sphero paddles in action.

Step 1: A Sphero-Sized Hole

Start by clicking the "Edit Grid" button in the lower-right corner, and change the unit to inches. Click "Update Grid" to finalize the settings. Next, click and drag a sphere shape from the "Geometric Shapes" menu to the workplane. Select the "Helpers" menu, and then choose

the "Ruler" tool. Click on the the box in the lower-left corner to reveal the dimensions of the sphere. Enter "3 inches" for the width, height, and depth of the sphere, as in Figure 11-25. When you are finished, click on the X to dismiss the "Ruler" tool. Select the sphere, and change the shape to a hole in the "Inspector" menu box. Now we have an idea of Sphero's exact size, and we can create our design around it.

Step 2: Ring Around a Sphero

To create the ring that will become our first paddle, click and drag the "Tube" to the workplane from the "Geometric Shapes" menu. Resize the tube using the "Ruler" tool to a width and depth of 4 inches, and adjust the height to ¼ inch (Figure 11-26). Center the tube vertically and horizontally on the sphere the best you can. Then select both shapes and click "Adjust" and then choose "Align." Align the shapes by clicking on the center alignment circles. You should have a shape that resembles the planet Saturn (Figure 11-27). Select both shapes and rotate them 90 degrees so that that the ring is

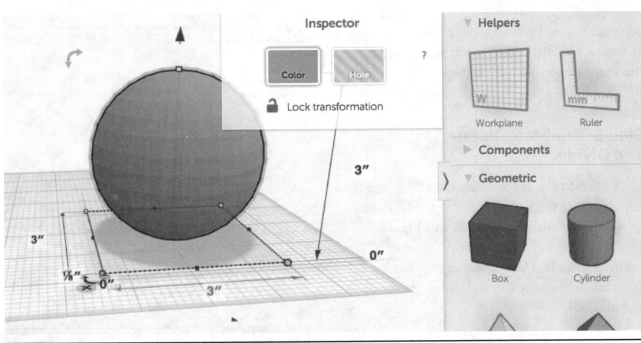

Figure 11-25 Adjusting sphere size with "Ruler" tool.

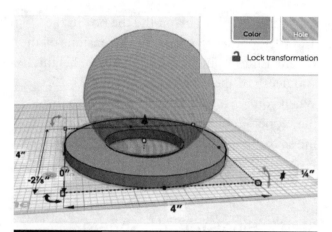

Figure 11-26 Resizing tube using ruler.

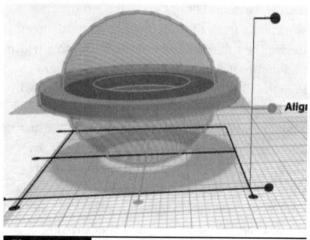

Figure 11-27 Aligning tube to center.

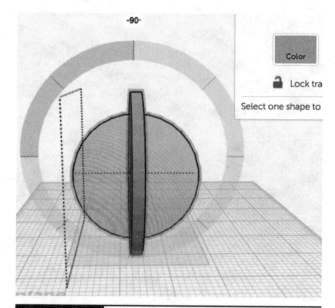

Figure 11-28 Rotating Saturn shape.

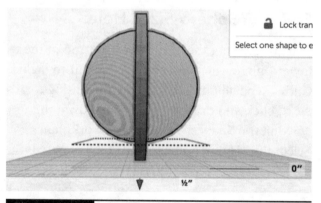

Figure 11-29 Lifting Saturn shape to workplane.

vertical (Figure 11-28). Click on the black arrow at the bottom of the Saturn shape, and move the shape up so that it is on the workplane, as in Figure 11-29.

Step 3: Duplicate, Rotate, and Divide

Click on the tube shape, select the "Edit" menu, and duplicate the first paddle you made. Rotate it to the left 90 degrees, as shown in Figure 11-30. Duplicate the original tube shape again and rotate it 90 degrees so that it is perpendicular to the original paddle, as in Figure 11-31.

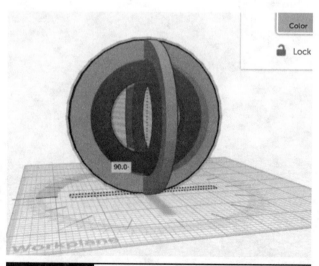

Figure 11-30 Rotating duplicate 90 degrees.

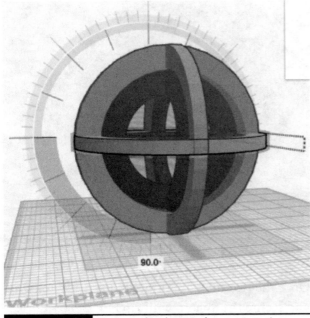

Figure 11-31 Perpendicular 90-degree rotation.

Now we are going to go back to the first paddle we made. Later we will divide this design in half, and this is a great opportunity to make that division precise. Click on the shape and use the white dot in the corner to change the width from ¼ to ⅛ inch, as in Figure 11-32. Select

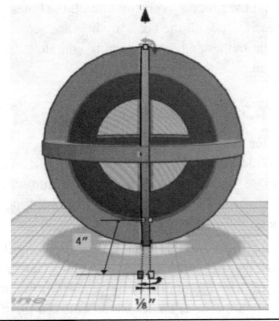

Figure 11-32 Resizing center vertical into ⅛-inch halves.

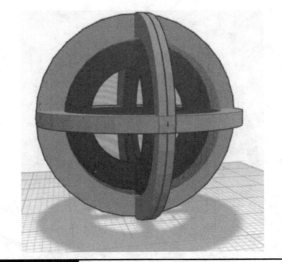

Figure 11-33 Two ⅛-inch paddles side by side.

the ⅛-inch-thick tube shape and duplicate it. Position the shape next to the other so that it looks just like it did before, only now you have a clear dividing line (Figure 11-33).

Step 4: Divide and Conquer

It is much easier to print two halves instead of a sphere. We also need a simple way that we can open the water paddle to place Sphero inside. Turn the workplane so that you are facing the edge of the original paddle and can see the two ⅛-inch panels. Make a large box hole about 5 inches tall, 2½ inches wide, and 5 inches deep. Because we have already divided the original paddle, simply align the box hole so that it covers half the shape, as shown in Figure 11-34. Click and drag to select both shapes, and select "Group." Press the "Group" button, and the final shape of the paddle will be revealed. Now that you have half the shape remaining, rotate it 90 degrees so that the flat side is facing down, and lower it to the workplane using the arrow shape that appears at the top when it is selected (Figure 11-35).

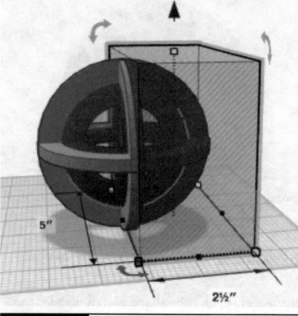

Figure 11-34 Box-shaped hole in position.

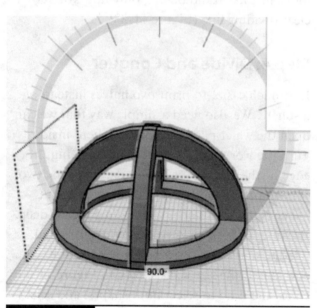

Figure 11-35 Half paddle rotated 90 degrees.

Step 5: Get It Back Together

Eventually, we will print two copies of the half paddle to make a whole. Once Sphero is placed inside the design, we need a way to put it back together. To do this, we are going to create four small holes and equally space them around the central ⅛-inch paddle as shown in Figure 11-36.

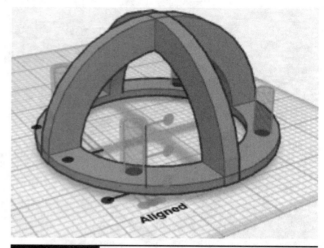

Figure 11-36 Aligning holes.

Once the two sides are printed, line these holes up and secure the Sphero with zip ties.

Start by dragging a cylinder-shaped hole to the workplane. Resize the hole to a diameter of ¼ inch. Click on the "Edit" menu and duplicate the hole three times. Position a hole in the center of each quarter of the central paddle, as shown in Figure 11-36. Select two of the holes, and click on the "Adjust" menu, then press "Align." Click on the alignment circles to ensure that the holes are in line. Rotate the paddle, and repeat the process to ensure that all the holes are aligned. Once all the holes are in place, select all the shapes and group them to complete the design.

Step 5: Print and Play

Center the finished half-paddle on the workplane, and then select the "Design" menu. Choose "Download for 3D printing," and select "OBJ" or "STL" depending on which file type you need for your printer.

Once both pieces are printed, you may need to use a craft knife to clean up rough edges. Cut two to three pieces of ⅛-inch-thick craft foam in 1-inch-diameter circles to rest between the inside of the paddle and the Sphero, as shown in Figure 11-37. The foam will help to keep Sphero

Figure 11-37 Foam inserts.

from moving around in the paddle and provide it with a good grip on the paddle. If you find that your Sphero can move inside the paddle, you may need to add another piece of foam.

Before you put Sphero in the paddle, be sure that it is fully charged. Line up the holes, and place the zip ties through them. Tighten the zip ties firmly, as shown in Figure 11-38, and trim them as needed. Remember to test your design and fit in a sink or tub and make a rescue plan before you set sail in a larger body of water!

Challenges

- How could you use Sphero to propel a boat?
- How could these basic paddles be improved to make Sphero work even better in the water?
- What other things could you design for Sphero?

"3D Printing" Challenge

Think about something at home that you could improve or fix with a plastic part. How could you improve the design of the part to make it stronger or work better?

Take pictures of your challenge project. Tweet it to @gravescolleen or @gravesdotaaron, tag us on Instagram, and include our hashtag #bigmakerbook to share your awesome creations on our community page, which will host photos of you and your maker projects.

Figure 11-38 Zip ties.

Mixing It All Together

THE LAST CHAPTER is full of makerspace mash-up fun. Now that you have filled your makerspace toolbox, it is time to try smashing maker tools together to invent all kinds of fun.

Project 48:	Adding a Makey Makey Go Switch and littleBits Audio to a Smart Phone Projector
Project 49:	littleBits Makey Makey Confetti Catapult Photo Finish
Project 50:	Musical Paper Circuits with Makey Makey and littleBits
Project 51:	littleBits Sphero Smart Track

Chapter 12 Challenge

"Makerspace Mash-up Fun" challenge. You've learned so much! Now, what will you invent and create?

Project 48: Adding a Makey Makey Go Switch and littleBits Audio to a Smart Phone Projector

Cost: Free–$$–$$$

Make time: 10–15 minutes

Supplies:

Materials	Description	Source
Previous project	Smart phone and cardboard projector	Chapter 3
Smart phone to female USB adapter	Micro-USB to female USB adapter (Android phones) Lightning to female USB adapter (iPhone)	Amazon
Makey Makey Go	Makey Makey Go, alligator test lead	Joylabz.com
littleBits	Power Bit (p1), microphone Bit (i21) with ⅛-inch male audio jack cable, slide dimmer (i5), speaker Bit (o24), 9-V battery	littlebits.cc

Step 1: littleBits Setup

Place your phone in the box in the location and orientation it will be used during projection. Cut a small hole in the bottom of the box, and slip the audio cable through. Plug the audio cable into the headphone jack and then plug in the opposite end into the microphone bit.

Step 2: Connect littleBits Circuit

Start by plugging the power bit into the microphone bit and then adding the slide dimmer. Attach the speaker to the microphone, and you are ready to power your circuit up.

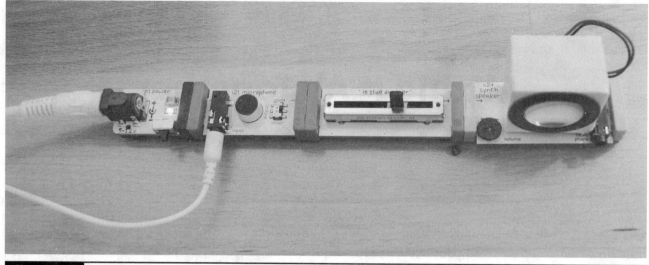

Figure 12-1 littleBits circuit.

Connect the battery cable to the power bit, and flip the switch ON. See the full littleBits circuit in Figure 12-1. Test your circuit by playing a video on your phone and adjusting the volume with the slide bit.

Step 3: Set Up "Play and Pause" Switch with Makey Makey Go

Leave your phone in place in the box, and attach the female USB to micro-USB adapter to your phone. These adapters are somewhat short, so you will need to make a hole as close as possible for this to work. Next, plug the Makey Makey Go into the other side of the USB adaptor. It needs to be outside the box so that the LED lights do not interfere with your projection. If you are using a metal table, put a placemat or thick piece of paper under the Makey Makey Go because you do not want it to incorrectly gauge the capacity because the metal table is stealing electrons. The Makey Makey Go will flash a series of colors when you plug it in to let you know that it is working (Figure 12-2).

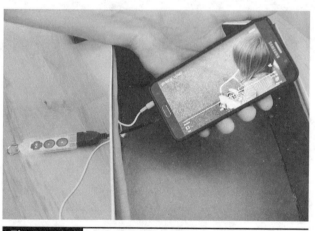

Figure 12-2 Attach audio cable and Makey Makey Go.

Step 4: Play and Pause with Pizza

Order a pizza! You are no doubt going to have greasy hands from eating pizza, and we don"t want to mess up our phone screen, do we? So we will add a Makey Makey Go to help us pause and start our movie, by sacrificing a small slice of pizza to function as our "Play" switch. If you are really hungry and you cannot spare a slice of pizza, I am sure you have a banana lying around you could use! After all, bananas and Makey Makey go hand in hand. Or you could try an olive or a house plant. All that matters is that the object you are inventing as your

"Play" switch has some type of conductivity to it. Fruit, cheese, bread, plants, pots, pans, and even people conduct electricity! Attach one end of the alligator clip to your object and the other to your Makey Makey Go. Remember that you can use anything that conducts electricity, so don't be afraid to experiment. When you plug in your Makey Makey, the end will flash a series of colors and then stop on blue, letting you know that it is ready to go.

After hooking up your object, hit the "Play" button to let Makey Makey Go know that it needs to recalibrate. When you do this, the Makey Makey dumps a ton of electrons onto the object to determine the object's electrical capacity. Then, when you touch the "Play" switch (your object), you add a ton of electrons to the object, which tells the Makey Makey to go, and it activates your switch! Cool, right? Think of it sort of like your smart phone power button. Your smart phone's screen works with capacitive sensing, but not until you turn it on. In the same way, the everyday object is able to hold electrons, but that capacity is not measured until you plug in the Makey Makey with an alligator clip (Figure 12-3).

You can tap the "Gear" button to set the Makey Makey Go to read "Mouse left click" (which lights the Makey Makey Go tail blue) or "Spacebar" (which lights the Makey Makey Go tail red.) You can even remap the spacebar on the Makey Makey Go to any computer key by going to makeymakey.com/remap. However, once you unplug the Makey Makey Go from a device, it will reset to spacebar.

If you hold the "Play" button, you'll put the Go into its sensitive setting. If you plug it into your pizza and it doesn't activate your phone, you may want to put your Go into sensitive mode (Figure 12-4). You'll know if it is sending a signal to your phone if the end turns green when you tap on the pizza.

Figure 12-3 Attaching pizza.

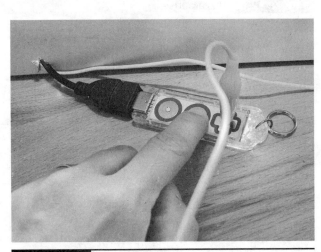

Figure 12-4 Recalibrating capacitive touch.

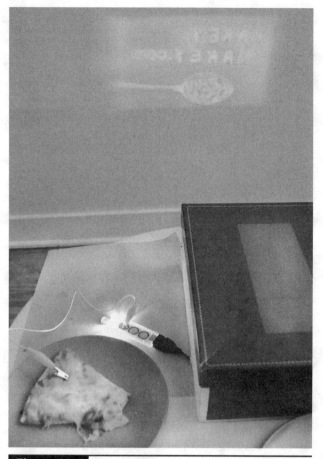

Figure 12-5 Pizza pause.

Cue up your favorite video on your phone, and don't forget to lock the orientation on your phone so that it will play upside down. Tap your object to start, and pause your video (Figure 12-5). This works great with videos from your video player on your phone and should even work with YouTube videos depending on your phone model.

Challenges

- We left our littleBits circuit on the outside of the projector box. How could you integrate it into the projection box?

- What other bits could you add to make this hack more amazing?

- What other things could you use for your Makey Makey Go switch?

- What other interactive things can you invent with your Makey Makey Go?

Project 49: littleBits Makey Makey Confetti Catapult Photo Finish

This project is versatile in that not only can it be triggered by conductive items, but it also can be used to hurl confetti, Ping-Pong balls, and a host of other materials. You can also use it to trigger an event on your computer. You can use a variety of boxes for this project as long as you can fit a plastic spoon onto the box with 1 inch to spare. The box used for this build was 5 × 7 × 10 inches.

Cost: Free–$$–$$$

Make time: 15–30 minutes

Supplies:

Materials	Description	Source
Previous project	Makey Makey pressure sensor	Project 38
Tape	Duct tape	Craft store
Recycled supplies	Cardboard box, plastic spoon, paper, tissue paper	Recycling bin
littleBits	Makey Makey Bit with included cables (w14), servo Bit (o11), power Bit, cable (o1), and 9-V battery	littleBits.cc

Step 1: Center the Servo

Turn the box on its side, and mark the center of the short sides. Draw a line down the middle of the side of the box. Position the servo on the short side of the box so that the servo spline is in line with the center. The servo should be high enough that the rectangular servo body is even with the top of the box. Place a mark on each side of the rectangular body. Use a screwdriver to poke holes through the box on these marks. Secure the servo to the side of the box with a zip tie (Figure 12-6).

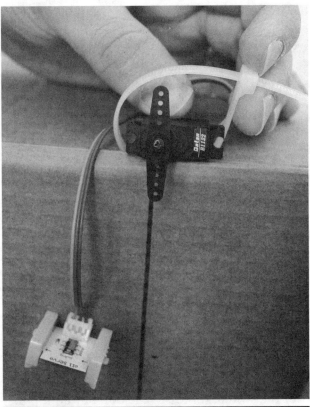

Figure 12-6 Mark, punch holes, and centerline and servo.

Rotate the servo all the way to the right. The servo arm needs to be in line with the center mark. If it is not, remove the screw and arm. Turn the arm so that it is in line with the center, and replace the screw. Set the switch on the o11 servo bit to swing.

Step 2: Position and Flex Throwing Arm

Plastic spoons have been used for years to flick peas at unsuspecting victims in the cafeteria. Their springiness and availability make them the perfect choice for a catapult arm. We especially like the superlong spoons you get with shakes and iced tea! Place the spoon on the center mark with half the bowl end under the servo. On the handle end, make a mark on the spoon about 1 inch away from the end. Remove the spoon, and draw a mark across the center. Use a box cutter to slice the box here. You might need to make

another slice about ⅛ inch away to get the spoon into the box (Figure 12-7).

Cut a 4-inch slice of duct tape, and place half on the box and half on the spoon (Figure 12-8). Slice the tape up from the end to the spoon on both sides, and then wrap the tape around the spoon (Figure 12-9). This will allow the spoon to move slightly side to side and easily back and forth.

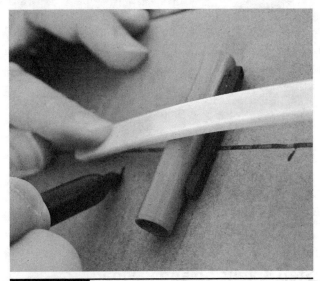

Figure 12-7 Marking spoon hole.

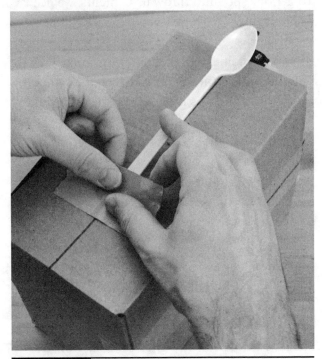

Figure 12-8 Tape half on spoon and half on box.

Figure 12-9 Tape cut and wrapped.

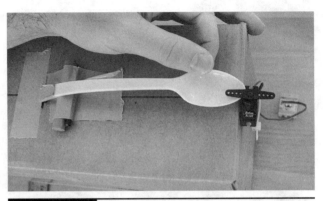

Figure 12-10 Under tension.

Step 3: Tension and Payload

To get our spoon to throw its contents, we need to put it under tension. An old marker lid or eraser works great for this task. The servo motor should be in line with the center. Its job will be to hold the end of the spoon and keep the catapult from firing until it is triggered by the w14 Makey Makey bit. Move the marker lid under the spoon, and try to find the spot that you can get the most tension and still get the end of the spoon under the servo without it breaking (Figure 12-10). Use a 3-inch piece of duct tape to secure the lid once you have it in the perfect spot.

A spoon is great for a pea, but when you want to throw a whole bunch of confetti, you need a larger bucket for your payload. The small side of an Easter egg or a small plastic cup works great. Try to keep it light, if possible, and secure the bucket with tape or hot glue. You can now test the catapult by rotating the servo by hand.

Step 4: Connect the Trigger Switch

We are going to use the Makey Makey littleBit to trigger the catapult to launch confetti at the racer. This celebration at the completion, a robot obstacle course will also launch a photo finish because we've hooked our MakeyMakey littleBit to the click on our computer. As the robot races over the switch, it presses click on our laptop to take a final race photo. You will need a pressure switch for the Makey Makey bit like the one featured in Chapter 8. Start by using an alligator clip to connect one side of the pressure switch to the Makey Makey littleBit on the click input. Connect another alligator clip to the opposite side of your pressure switch and the other end to the earth input on the Makey Makey littleBit. Move the pressure switch onto the center of your track.

Connect the w14 Makey Makey bit to the o11 servo bit. Place a p1 power Bit to the center of the Makey Makey bit, and turn it on (Figure 12-11). Load the bucket with a payload of confetti, and test the pressure switch by pressing on it. What happens (Figure 12-12)?

Step 5: Photo Finish

If you have a laptop with a camera, you can open up a photo app such as Photobooth to take pictures. The confetti catapult is a great match because you can catch reactions or record

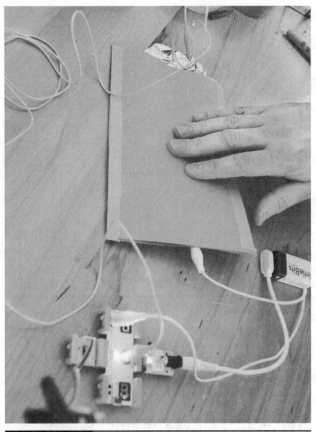

Figure 12-11 Complete circuit.

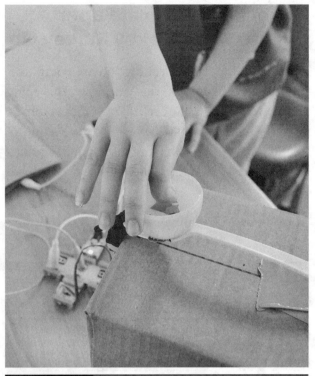

Figure 12-12 Payload of confetti.

a photo finish on your obstacle course. Connect the Makey Makey bit to your computer; use the included micro-USB cable. Hover the mouse over the "Take a Photo" button, and then activate the pressure switch to take a photo and launch some confetti (Figure 12-13)!

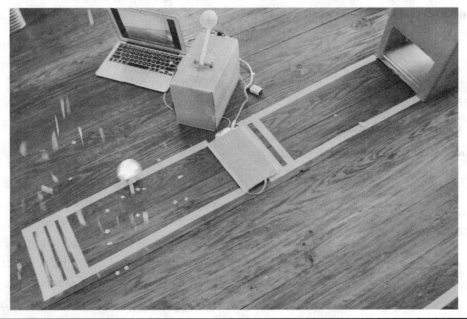

Figure 12-13 Photo confetti finish.

Project 50: Musical Paper Circuits with Makey Makey Go

If you didn't make Project 21, you are going to want to go back and do that one now! This epic Makey Makey Go hack relies on your awesome circuitry project from Chapter 5. We are going to add music to this blinking disco paper circuit and get the party started!

Cost: $–$$

Make time: 15–30 minutes

Supplies:

Materials	Description	Source
Previous project	Disco paper circuitry	Project 21
Phone or computer	Computer with USB or smart phone with USB/ micro-USB adapter	Amazon
Makey Makey Go	Makey Makey Go, alligator test lead	Joylabz.com

Step 1: Gather Supplies

Get out Project 21, and dust off that paper circuit. Now let's add some beats to an already awesome lightshow. Grab your Go, some aluminum foil, and an alligator clip. Cut a small piece of foil, but keep a larger strip handy as well. We are going to do a little capacitive experimenting.

Step 2: Attach Go

Plug your Makey Makey Go into a USB port on your computer. The lights should flash and then stay blue. This means that the Go is set to "right-click." To change the Go to "spacebar," just tap the setting wheel, and the end of the Makey Makey Go should turn red. Now grab an alligator clip, and hook one end to the + sensor

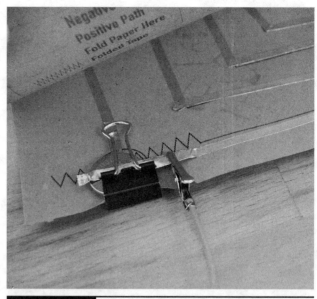

Figure 12-14   Alligator clip to copper flap.

on the Go and the other to the copper flap on your paper circuit, as in Figure 12-14.

Step 3: Test and Experiment

First, try tapping the tape to see if you will make the Makey Makey Go activate. Now try closing your paper circuit and tapping the tape from the outside. Doesn't work? Why? Because the paper is nonconductive, and we are no longer adding to the capacity of the copper tape.

You don't have to use a human to activate your Makey Makey Go! You can always use another conductive object to increase the capacity on the object hooked to your Makey Makey Go clip. In this project, since we are using the line of copper tape as our Makey Makey Go switch, we can easily add some aluminum foil to the top inside of our paper circuit so that when we touch the card from the top, the foil will touch the copper tape, and we can turn on the music. Just like the pizza switch, when the Makey Makey Go is hooked to the copper tape and we tap the "Play" button to recalibrate, the Go reads how capacitive our tape tracing is. So, if we add another piece of aluminum foil to the top of our

Figure 12-15 Test.

card, when it touches the copper tape, it changes the electron capacity of the copper tape, thus activating the Makey Makey Go. Try a small piece of foil and a larger piece of foil? What happens (Figure 12-15)?

Step 4: Add Conductor

When you find the size foil you want to use, put some double-sided tape on one side, and lay it over the copper tape to ensure correct placement. Then close your paper circuit card to adhere the foil to the top of the card, as in Figure 12-16.

Step 5: Ready to Play!

Cue up your favorite YouTube video or soundtrack, tap your Go switch to start the music, and slide your finger to blink the disco lights! You can stop and start the music by tapping the top of your paper. If the light on the end of the Makey Makey Go turns green, then you are activating the switch (Figure 12-17).

Figure 12-16 Correct placement trick.

Figure 12-17   Green means go!

Figure 12-18   Setting to spacebar (red).

Step 6: Troubleshoot

If your music isn't starting, check to see if you have your Go set to spacebar or right click. Tap the gear to set the Go to spacebar, and the end will turn red, as in Figure 12-18. Tap again to set it back to right click, and the end will turn back to blue. If you tap the paper where you made your switch and the tail turns green, you are activating the switch (Figure 12-19). If you have it plugged into your device and this doesn't work, check to make sure that you have the correct setting for your device. On a computer, spacebar works great for YouTube!

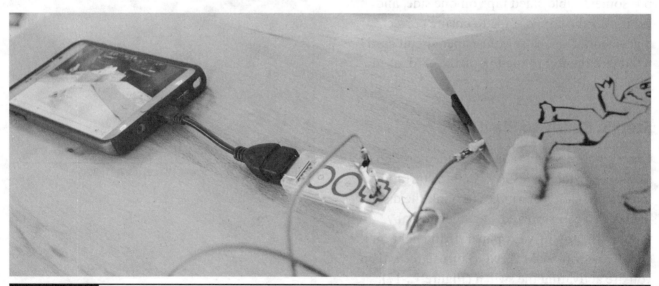

Figure 12-19   Testing for green!

Project 51: littleBits Sphero Smart Track

It is time to harness the maneuverability of Sphero and hone your programming skills with a smart track. You will need some obstacles, and you can pick and choose from the previous projects listed below or make new ones.

Cost: $–$$$

Make time: 15–30 minutes

Supplies:

Materials	Description	Source
Track supplies	Low-adhesive masking tape Graph paper Pencil	Office supply store Hardware store
Track tools	T square (optional) Measuring tape (optional)	Hardware store
Previous project	Arduino littleBits project	Project 21
Previous project	Robot arms and moving gates project	Project 41
Previous project	Flashing rainbow lamp or tunnel	Project 42
Previous project	Classic miniature golf windmill	Project 43

Step 1: Obstacle-Course Philosophy

Think about the strengths and limitations of Sphero. For example, Sphero can't go from a complete stop and then make a jump 2 inches away. Also, consider what makes an obstacle course fun? How can you make your obstacle course challenging to program but not impossible? How can you make it fun and functional?

Step 2: Draw It Out and Test Obstacles

Once you have thought through the obstacles you want to use, it is time to plan your course. Designing your course on graph paper works well because you can use 1-inch scale to calculate your track in feet. You may need to test obstacles that you have made in order to calculate the amount of distance required for Sphero to successfully navigate them. We recommend a track that is at least 1 inch wide, except for places where you are creating an obstacle with width as part of the challenge.

Step 4: Arduino littleBits Tunnel

Bring out your Arduino littleBits project from Chapter 5, and adjust your lights (Figure 12-20) to make a beautiful rainbow tunnel. You can extend the lights by using the wire Bit, as shown in Figure 12-21. Create a channel for your

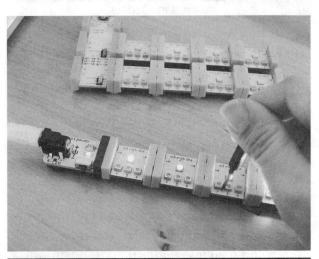

Figure 12-20 Adjusting lights.

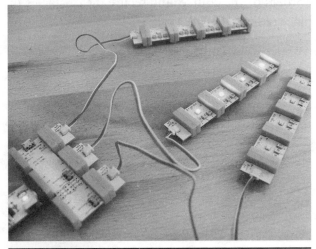

Figure 12-21 Extend with wire Bit.

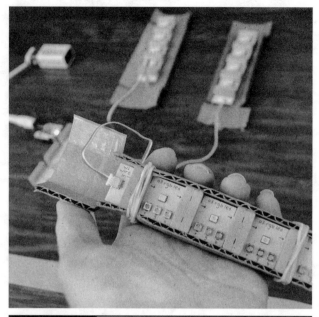

Figure 12-22 littleBits in channel.

littleBit lightshow out of cardboard as you did in Chapter 11 (Figure 12-22).

Use a zip tie to attach the Arduino Bit and power Bit to the outside of the box, as in Figure 12-23.

Step 3: Lay Out and Test Drive

Place all your smart obstacles, and lay out the track with small pieces of tape. Test drive the course, and then make adjustments. When you have a drivable track, lay down tape along the borders of the course, and secure the obstacles in place (Figure 12-24).

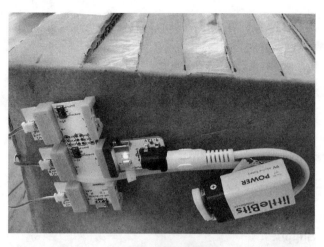

Figure 12-23 Attach bits.

Step 5: Program and Challenge

Use the Tickle app or the SPRK Lightning Lab app to write a program for your obstacle course! It's a great time to compete with others for fastest time and most accurate programming (Figure 12-25).

Classroom tip: This is a great time to go through the design process with participants. It is helpful to complete some littleBits projects ahead of time so that everyone is familiar with the functions and capabilities of littleBits.

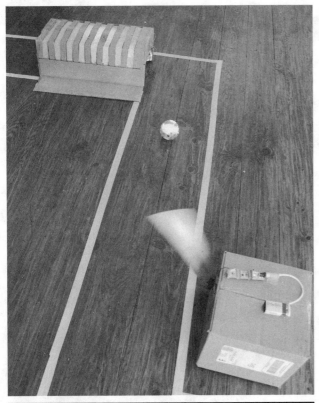

Figure 12-24 Making adjustments and driving track.

"Makerspace Mash-Up Fun" Challenge

Wow! What a journey! Think about all the things you've learned to make and create! Now it's time to think about other things you could put together. How could you use the littleBits with your diddley bow? Could you mash up your brush bots with Makey Makey? Could you program robots to activate Makey Makey switches? Could you take apart toys and rebuild them with littleBits?

We hope this book has filled your maker toolbox with many new skills and you are ready to use all of the knowledge you gained as building blocks for designing and making your own awesome creations! It's time to start experimenting by designing, making, tinkering, and thinking up your own projects. We can't wait to see what you create!

Figure 12-25 Running a program on the track.

Templates for Paper Circuits and Sewing Circuits

THESE PAPER CIRCUIT templates are here to help you make your own cards for combining the awesome power of low tech crafts with high tech circuits. Use them to enjoy making your own electronic cards for yourself or with students. Follow the projects in Chapter 4.

The sewing circuit templates go alongside the projects in Chapter 7. We are providing you with full circuitry details and a blank template for you guitar if you want to design your own circuitry.

The Big Book of
MAKER SPACE Projects

Inspiring Makers to Experiment,
Create, and Learn

Colleen Graves and Aaron Graves

Project courtesy of

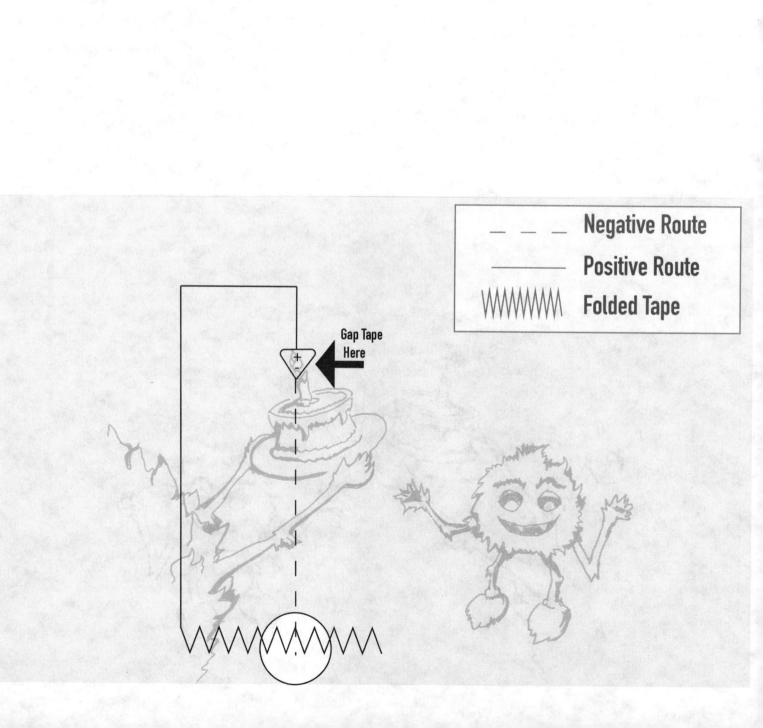

The Big Book of
MAKER SPACE
Projects

Inspiring Makers to Experiment, Create, and Learn

Colleen Graves and Aaron Graves

**Folded Copper
Tape Flap Switch**

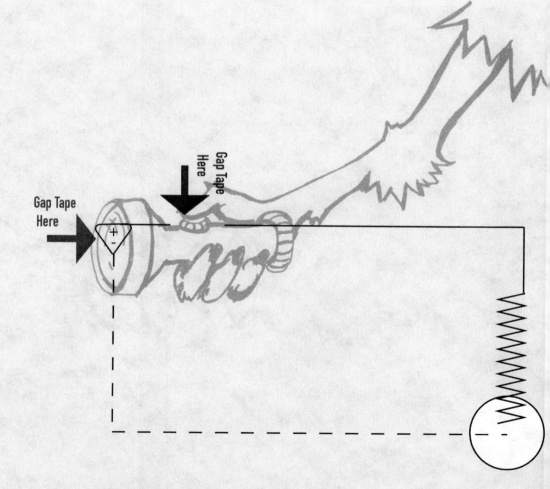

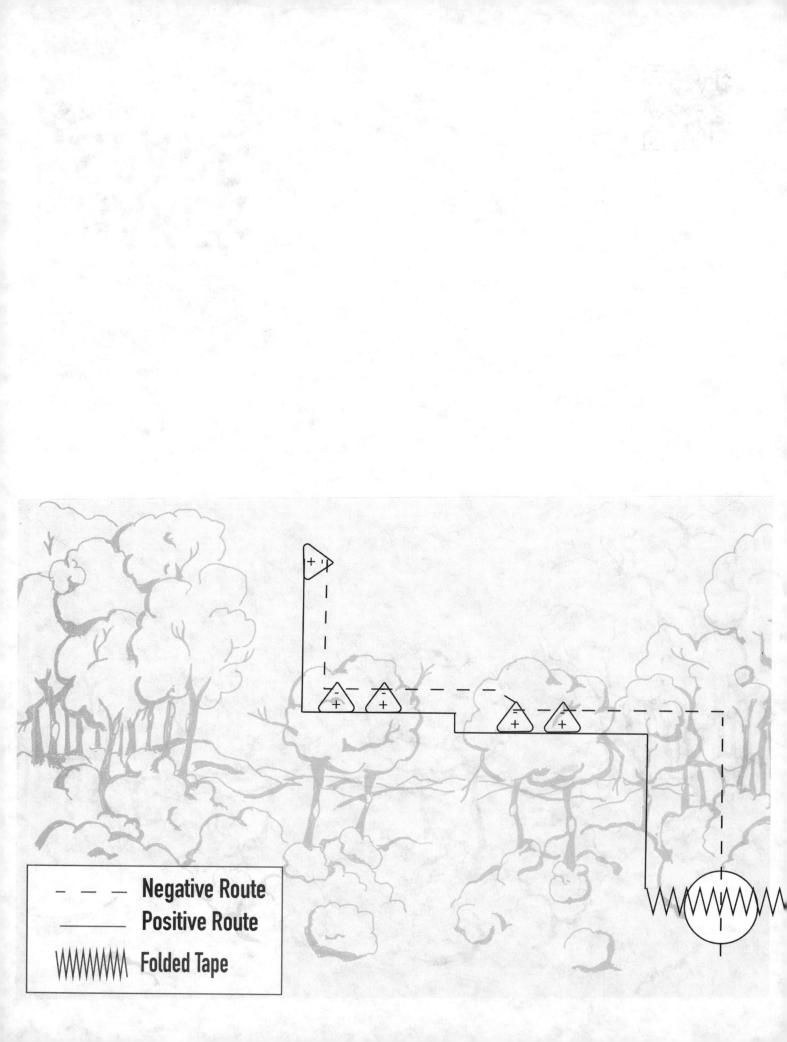

Negative Route

Positive Route

Folded Tape

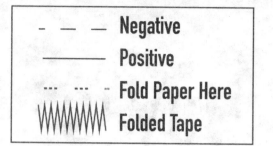

Negative
Positive
Fold Paper Here
Folded Tape

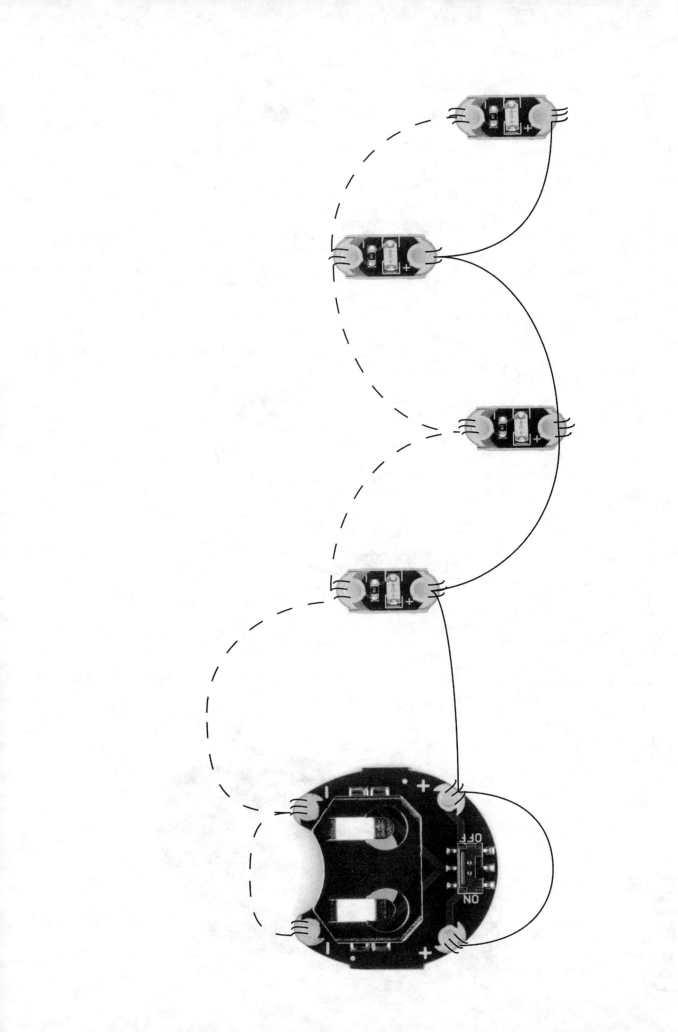

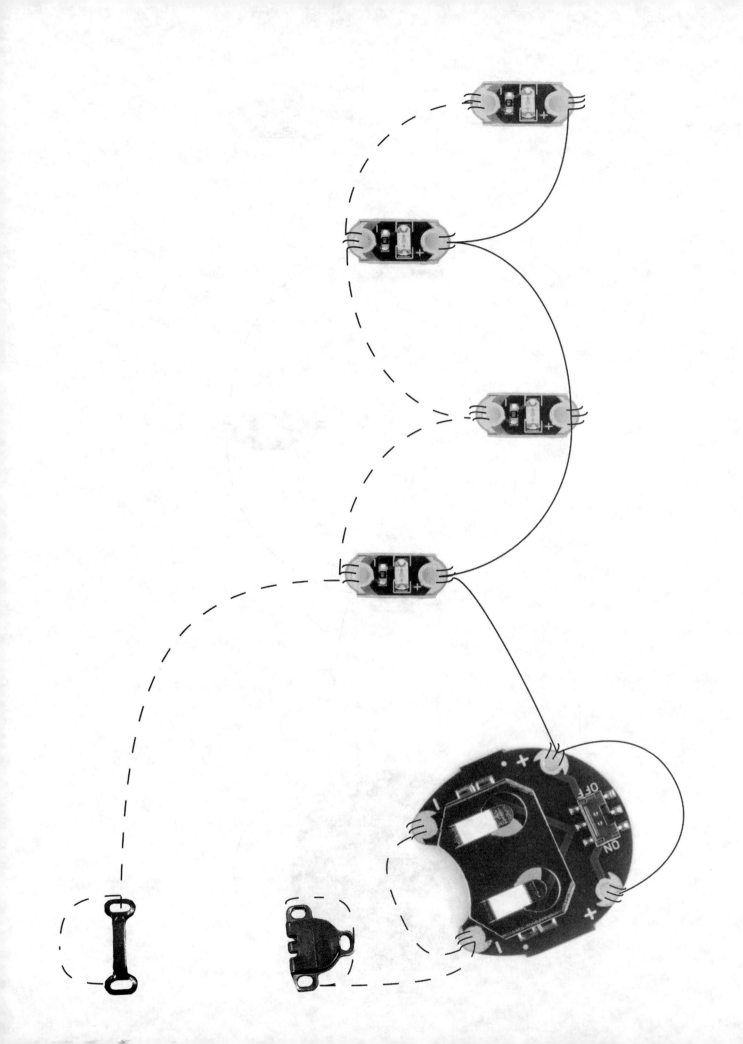

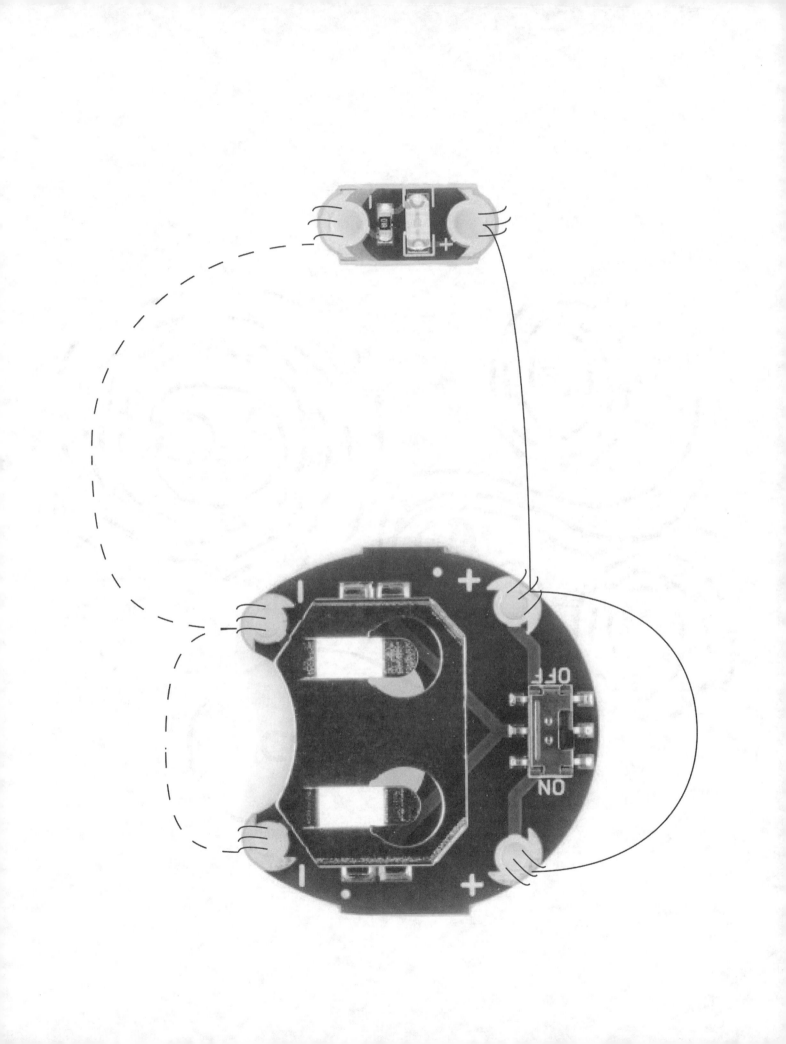

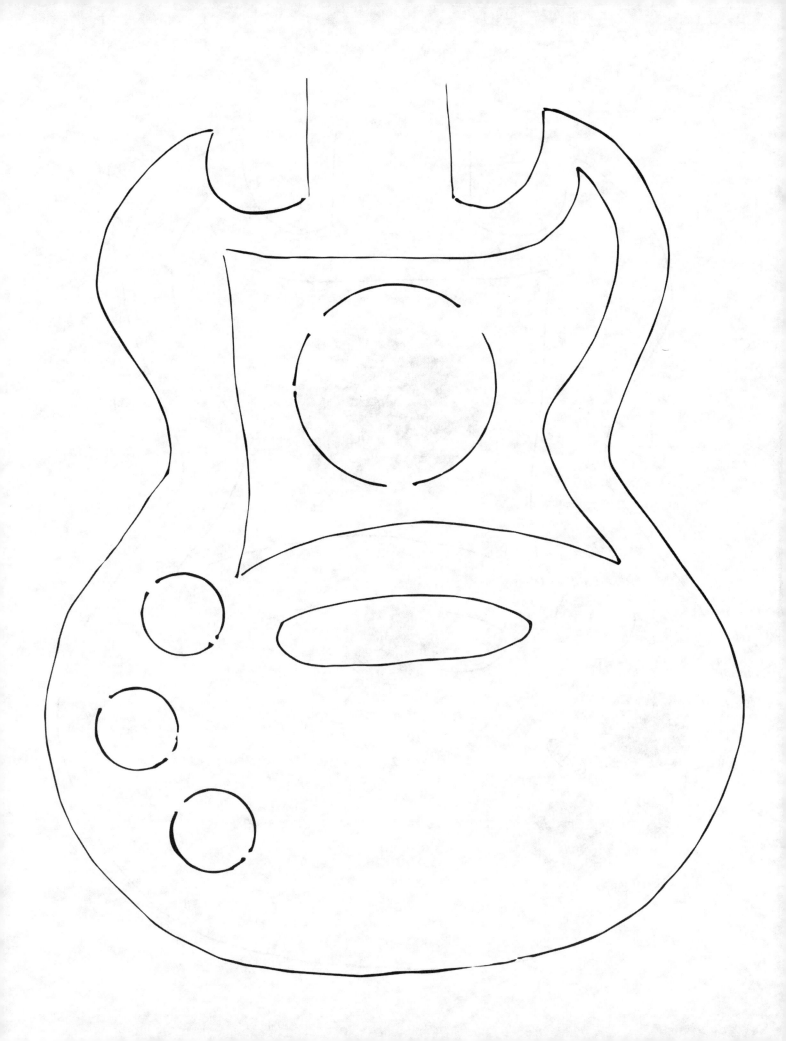

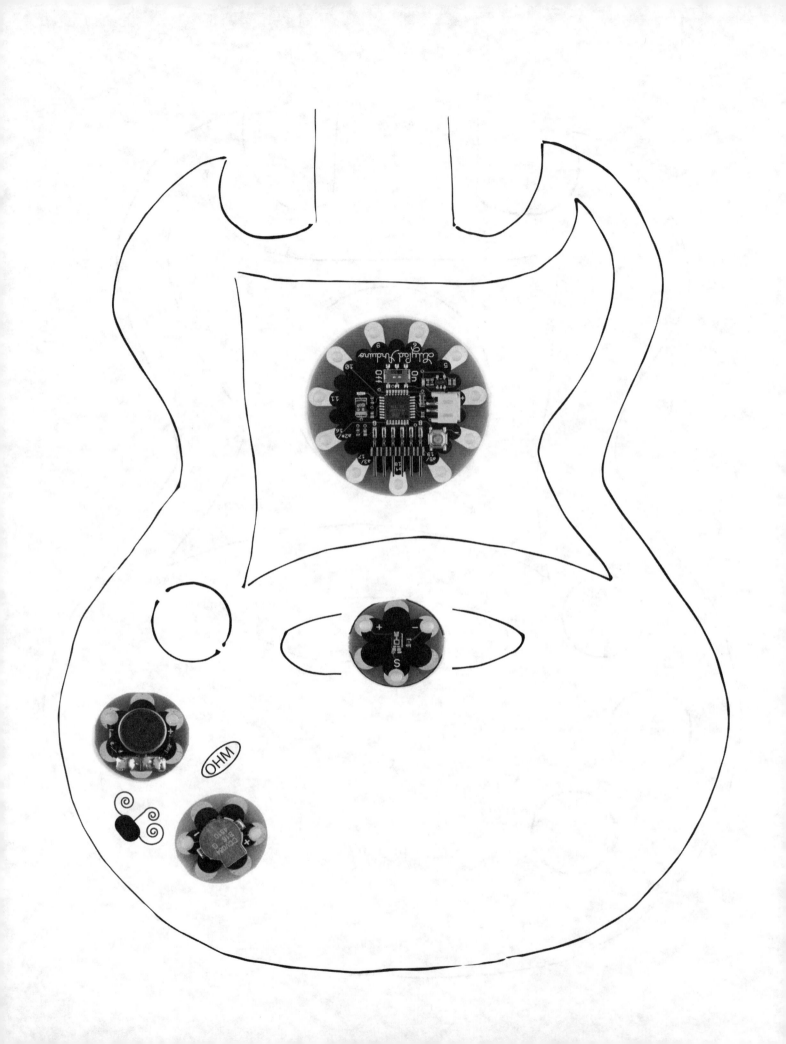

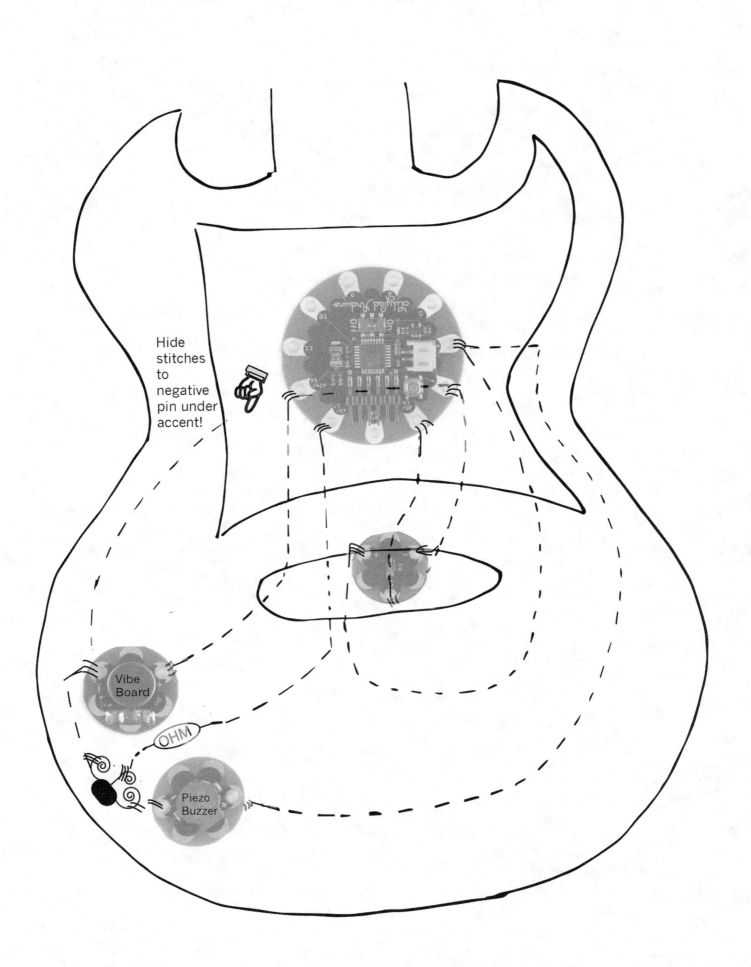

Hide
stitches
to
negative
pin under
accent!

Vibe
Board

OHM

Piezo
Buzzer

Index

W